Animal Athletes

Animal Athletes

An Ecological and Evolutionary Approach

Duncan J. Irschick

Professor, Department of Biology, University of Massachusetts Amherst, USA

Timothy E. Higham

Associate Professor, Department of Biology, University of California, Riverside, USA

OXFORD
UNIVERSITY PRESS

OXFORD
UNIVERSITY PRESS

Great Clarendon Street, Oxford, OX2 6DP,
United Kingdom

Oxford University Press is a department of the University of Oxford.
It furthers the University's objective of excellence in research, scholarship,
and education by publishing worldwide. Oxford is a registered trade mark of
Oxford University Press in the UK and in certain other countries

First Edition published in 2016

Impression: 1

Published in the United States of America by Oxford University Press
198 Madison Avenue, New York, NY 10016, United States of America

British Library Cataloguing in Publication Data
Data available

Library of Congress Control Number: 2015944343

ISBN 978–0–19–929654–5 (hbk.)
ISBN 978–0–19–929655–2 (pbk.)

Printed and bound by
CPI Group (UK) Ltd, Croydon, CR0 4YY

Foreword

This book examines how animal athletes have evolved. While many of us might normally think of the activities associated with "athletes" as consisting of only running and jumping, we set out to show that animals perform many tasks to an amazing degree of accomplishment. Some of the examples might seem obscure at first—for example, how can the rattling tail of a rattlesnake be considered an "athletic feat"? Once one realizes that the snakes vibrate their tail at 90 Hz for over an hour, then one's view changes. If you don't believe us, try shaking your hand as quickly as you can for an hour! In other words, the mundane world of animals is filled with extraordinary accomplishments, and our book sets out to celebrate this wonderful diversity by focusing not just on spectacular feats of running and jumping but also on feeding, vocalization, diving, flying, and many other feats. Each of these facets operates in a larger evolutionary and ecological context, and therefore we wanted to examine not only the "how" of amazing animal athletic feats but also the "why." Why do seals dive so deep? Why do some lizards live on twigs? Why do frogs vocalize for long periods of time when, in so doing, they expend a tremendous amount of energy? Our belief is that only in the broader context of ecology and evolution can these questions be answered and the broader pattern of animal athletic traits be understood.

Therefore we delve into many areas—such as, for example, how natural selection and sexual selection operate on performance traits, the musculoskeletal basis of these traits, and how such traits vary in different geographical or evolutionary settings. Our examples are wide-ranging and are drawn from a range of invertebrates and vertebrates, including frogs, lizards, sharks, rodents, bats, birds, insects, spiders, and more! Our goal was not to be comprehensive in our analysis of different systems, and therefore our approach is exemplary rather than exhaustive. We took this approach to enhance readability, and readers are recommended to read many of the cited papers and books to learn more on a certain topic. Our language was intended to be simple, and therefore not every study is covered in the exhaustive detail one might expect to find in a typical academic paper. Our book was hopefully written so that many could access it, but we include enough detail that (we hope) will please specialists. We sincerely hope that this book inspires a new way of thinking about animal athletics and spurs new research into the integrative biology of complex functional traits.

This book could not have been written without the assistance of many individuals. A number of individuals read key chapters and provided invaluable assistance. These individuals include Anthony Herrel, Scott Kelly, Rodger Kram, Henry Astley, Kiisa Nishikawa, William Hopkins, Jerry Husak, Jonathan Losos, Daniel Moen, Jennifer Grindstaff, Simon Lailvaux, Sheila Patek, Raoul Van Damme, Roberto Nespolo, and others. Several discussions with members of the Irschick and Higham lab groups solidified some ideas regarding performance as well as helped identify idiosyncrasies that frequently arise in science. Our editors Lucy Nash and Ian Sherman were especially helpful in guiding us throughout the book-writing process by keeping us on schedule and providing invaluable assistance in the editing process. Their patience during the ten (!) years of writing and during our many twists and turns is greatly appreciated. Finally, we could not have written this book without the support of our families. Specifically, Duncan Irschick thanks Jitnapa Suthikant, as well as Darwin and Calder Irschick for their support and patience. Tim Higham would like to thank Melissa, Daphne, Iris, and Violet for their encouragement, enthusiasm, and endless support.

Contents

1 Animal performance: An overview 1

1.1 Why study performance? 1
1.2 Definitions of performance 6
1.3 The hierarchical nature of biological systems 10
1.4 Variability, repeatability, and heritability 11
1.5 The ecological context of performance 13
1.6 The evolution of performance 15
1.7 Behavior and performance 17
1.8 Mechanistic and energetic constraints on performance 21

2 The ecology of performance I: Studies of fitness 24

2.1 Natural selection, sexual selection, and performance 24
2.2 Path diagrams of selection 25
2.3 Definitions of fitness 27
2.4 Quantifying the impact of natural and sexual selection 28
2.5 Does high performance result in higher fitness? 29
2.6 Is selection on morphology stronger than for performance capacity? 35
2.7 Manipulative studies of animal performance and selection 38
2.8 Complex forms of selection on performance 43

3 The ecology of performance II: Performance in nature 47

3.1 Why study animal performance in nature? 47
3.2 What is ecological performance? 49

3.3   Impact of behavior on ecological performance   50
3.4   Environmental impacts on ecological performance   53
3.5   What percentage of maximum capacity do animals use in nature?   59
3.6   Emergent behaviors in the field: Defying physiological constraints   61
3.7   Slackers and overachievers: Behavioral compensation in nature   62

4   The ecology of performance III: Physiological ecology   67
4.1   Environmental influences on performance   67
4.2   Energetics and movement   68
4.3   Aerobic versus anaerobic metabolism   71
4.4   Born to run: Variation among animal species in the energetics of movement   73
4.5   Feel the burn: Harmful byproducts of movement   76
4.6   Temperature and animal performance   78
4.7   Acclimation and temperature   82
4.8   The ecological context of temperature   85

5   The evolution of performance I: Mechanism and anatomy   91
5.1   Mechanistic limits on performance   91
5.2   The mechanism and anatomy of trade-offs   92
5.3   Variation in anatomical structure as a foundation of animal performance   96
5.4   Creative arrangement of anatomical structure to enable high performance in suboptimal settings   99
5.5   Anatomical structure as a way of overcoming physical limits   103
5.6   Highly specialized performance: How do they do that?   108
5.7   Elastic mechanisms for powering rapid movements   110

6   The evolution of performance II: Convergence, key innovations, and adaptation   117
6.1   How does performance evolve?   117
6.2   Quantitative analysis of evolutionary data   119
6.3   A model for understanding the evolution of performance   120
6.4   Convergent evolution of morphology and performance traits   120

6.5   Tinkering and animal performance: How variation among species has resulted in evolution of novel performance capacities   130

6.6   Key innovations, adaptive radiation, and the invasion of novel habitats   135

7   Trade-offs and constraints on performance   143

7.1   Trade-offs, performance, and optimization   143

7.2   Why expect a trade-off?   144

7.3   Trade-offs within and among species   146

7.4   Mechanical constraints on performance   147

7.5   Innovations, constraints, and trade-offs   155

7.6   Ecological and reproductive constraints on performance   158

7.7   Overcoming trade-offs: The role of behavior   160

8   Sexual selection and performance   163

8.1   What is sexual selection?   163

8.2   A functional approach to sexual selection   164

8.3   Male competition and performance   165

8.4   Are sexual signals honest signals of male performance?   167

8.5   Sexual selection imposing costs on performance   171

8.6   Female choice and performance   174

8.7   Asymmetry and performance   177

9   Extreme performance: The good, the bad, and the extremely rapid   181

9.1   Extreme performance: If the Olympics were open to oribatid mites   181

9.2   Extreme performance: Overcoming limits   182

9.3   Need for speed: Extremely rapid movements   184

9.4   The evolution of morphological novelties   190

9.5   Morphological and physiological mechanisms of extreme performance   192

9.6   Behavioral means for greatly enhancing performance   195

10   Genetics, geographic variation, and community ecology   200

10.1   Animal athletics from a population perspective   200

10.2   The genetic basis of animal athletics   202

10.3   Geographic variation in performance capacity   205

10.4 The role of community structure in molding animal athletics   208

10.5 Many-to-one mapping, and communities   214

11 Human athletics: A link to nonhuman animals?   220

11.1 Humans versus other animals   220

11.2 Performance specialists versus generalists   221

11.3 Training impacts   222

11.4 Steroids and performance   226

11.5 Are athletes anatomically different?   229

11.6 Extreme sports and human mortality   231

12 Conclusion   236

Artwork credit lines   239

Index   243

1 | Animal performance
An overview

1.1 Why study performance?

What is performance ability? If one were to ask this question in any classroom or lecture hall (and we have!), one would immediately be greeted by a showing of hands, followed by a variety of different opinions. In the common vernacular, good "performance" can mean almost anything, such as how well one dances, sings, or even earns money. Simply put, "performance" can mean many things to many people, but a concept is only as useful as it is specific and explanatory. An overly broad use of the term "performance" is meaningless unless it allows us to examine a common body of knowledge. Fortunately, in the world of functional morphology and evolution, this term has enjoyed a more specific meaning that we will focus on throughout this book: performance capacity is any quantitative measure of how well an organism performs an ecologically relevant task that is vital for survival. As explained below, there are some more stringent assumptions behind this definition, but let's put that aside for now. Some classic examples include how fast an animal can run, jump, bite, fly, or perform nearly any athletic feat. You could even think of animal performance as being analogous to the human Olympics, although in the nonhuman animal world, the consequences of a bad day at the track are far more dire than any for even the most heartbroken athlete who has lost his or her event. Indeed, one of the hallmarks of animal performance is that nonhuman animals can place themselves at tremendous risk when performing "animal Olympics"; a cheetah or lion running after its prey can have its jaw dislocated if it miscalculates during the closing phase, and a miscalculated bite on a hard object can irreversibly damage a dog's jaws. One could speculate about the level of performance we might see at the human Olympics if lions were chasing those sprinters!

A study we had each heard of when we were graduate students crystallizes the difference between nonhuman animal and human performance. In a study in the 1980s, the physiologist Todd Gleeson performed a simple experiment with two groups of fence lizards (*Sceloporus*; a lizard commonly seen basking on trees, fences, and rocks in the western United States). For the "control" group, he allowed the lizards to stay in their small cages for about 6 weeks (Gleeson 1979). For the "experimental" group, he induced them to perform aerobic exercise on a treadmill in a training regimen for the same length of time. After 6 weeks had passed, he compared the aerobic capacities of both groups (the control and experimental) and found, much to his surprise, that the two groups did not differ! Obviously, if one performed the same experiments with humans, one would find that the group that had been trained would not only have superior aerobic capacities compared to the control group but also possess several physiological differences, such as increased capillary networks adjacent to large skeletal muscles, increased maximum aerobic capacity, improved blood pressure, and so forth. While not every animal shows this lack of a response to exercise (see chapter 11), for those species that do, it signifies the profound divide between human and nonhuman animal physiology. *Sceloporus* lizards rely on both burst speed and aerobic capacity to elude predators, capture prey, and defend territories, and the relentless pressure of natural selection over millions of years has not only removed many of the "weakest" individuals, it has done so to such an extent that natural selection has a far more limited menu to choose from. In other words, performance capacity is life and death for animals and has shaped their morphology, behavior, and physiology over vast stretches of evolutionary time.

The concept of performance also holds a key place in the human consciousness, both in terms of our own lives and for the natural world around us. Much of our obsession with performance dates to Charles Darwin and his synthetic theory of natural selection. A central thrust of his theory of natural selection is that some organisms will be more fit than others, and one of the main reasons for this variation in fitness is the ability of animals to perform such feats as running, jumping, biting, and so forth. Consequently, one implication of his theory is that some organisms are better performers than others and that this variation is vitally important to survival. His theory has also been extrapolated to include social aspects of human behavior—"Darwinism" or the "survival of the fittest," the idea that only the strong survive—an appealing, though somewhat misguided, view of human culture. Nonetheless, the idea of performance capacity as an arbiter of who lives or dies, or reproduces or not, has percolated in our collective consciousness and has been a spur for biological research into the process of adaptation.

We are fascinated with the traces that, in our everyday lives, reflect our obsession with performance. Each year, people lose millions of dollars betting on which dog or horse will be the winner at the racetrack, and entire countries have been known to explosively celebrate or fall into despair at the success or defeat of their favorite athletes at the Olympics. It is performance that drives almost 200,000 people to attend a single football (termed "soccer" in North America) match in Brazil. At such events, one can clearly see the range of performance abilities in nonhuman animals and humans; while some horses or humans are superior runners, others lag behind. But the artificial world of nonhuman animal and human athletics is not the only window into the mysteries of animal performance, as the natural world offers many strange and wonderful examples of the extremes to which animals have evolved various kinds of performance traits. Even the most jaded observer is fascinated when watching a small frog, only as small as a thumbnail, emit a piercing call that echoes over great distances. Most of us marvel at the rapid speeds of cheetahs when watching them chase their prey on the African plains. We are often caught pointing up in the air as hummingbirds zip by on a nice spring day, saying, "Wow, look at that!" Even in the mundane setting of our homes, the powerful crunch of a dog's jaw on a bone is a reminder of the extreme biting capacities of man's best friend. Evolution has seemingly endowed animals with amazing abilities that can exceed the limitations of animal morphology.

This book examines the broad scope of animal performance and integrates information on morphology, behavior, ecology, and evolution to understand the significance of performance in the lives of animals. We have two primary goals: first, to provide a perspective on the range of performance capacities present in nature; and, second, to understand the ecological and evolutionary context in which these performance capacities have evolved. To address these broad goals, we will explore how five factors, namely, behavioral, ecological, evolutionary, morphological, and physiological, act as innovators and constraints for generating diversity in animal performance. These factors allow us to identify how different parts of an animal work in concert to execute behaviors. It is during this behavior that we can define the kind of performance we are interested in. One theme we will explore is the underlying complexity of animal performance. One cannot fully comprehend the diversity of animal performance either by reducing an animal to its parts (reductionism) or by focusing solely on each factor (behavioral, ecological, evolutionary, morphological, and physiological) without considering interactions among these parts and how the whole organism behaves in nature. This integrative approach is important because performance capacities emerge from the whole organism, not from individual parts, and it is the whole animal that lives and dies and that passes on its genes (or not) to the next generation.

On the other hand, there is much to be learned by studying a few components in detail; but the important point is that reductionist studies are only pertinent when we also understand how the whole organism functions.

Beyond the aesthetic pleasure one derives from watching animals run, swim, jump, or fly, why should biologists study animal performance and how will such a focus provide insights into the broader fields of evolution and ecology? To address this, let's consider how natural selection works. Within any animal population over a span of time, individuals will either live or die, and some of those individuals will reproduce, while others will not. For the most part, factors that cause death or that influence the probability of reproduction, such as temperature, food, predators, and so forth, act on the whole organism. Let's consider an example that explains this point. Imagine a male frog calling on a summer night. The main reason that male frogs call is to attract mates. For most frogs, males that call the loudest and the longest are most likely to attract female frogs (Wells and Taigen 1989) and hence are more likely to reproduce and increase their overall fitness. Based on this line of logic, it seems reasonable that every male should have evolved the ability to call both loudly and for long periods; however, an inspection of both variables (call loudness and call duration) shows this not to be the case. As with many biological systems, there is tremendous variation in both traits, with some males being able to call both louder and longer than other males, often by large margins. The reason for this inequality is that calling in frogs is energetically expensive (fig. 1.1) and requires substantial aerobic investment, which is driven by muscles that are specialized for long-duration aerobic movements (Wells and Taigen 1989). Further, frog vocalization is a complex process, involving both a variety of muscle groups and complex behaviors, such as extension of the throat sac. However, because females choose males on the basis of call loudness and call duration, natural selection will operate *primarily* on these aspects of male performance (in this case, a specific form of natural selection called "sexual selection") and only *secondarily* on other aspects of this complex system. In other words, one can state with certainty that "natural selection favors frogs with large aerobic muscles" for long bouts of calling, but in fact selection favors individual frogs that can call for long durations. Female frogs care nothing for the muscles involved in calling. That feature is favored by natural selection only because it is necessary for frogs to call for long periods of time. In short, performance abilities are the "face" that animals put forth in the natural world, with many other supporting features "hidden" in the organism (although in some cases, natural selection does act directly on morphology or behavior alone). This example underscores how one cannot understand the evolution of unique morphological traits (i.e., the throat sac and the specialized muscle groups) or behavior (i.e., vocalization) without understanding the resultant performance capacity (call intensity and duration) or vice versa.

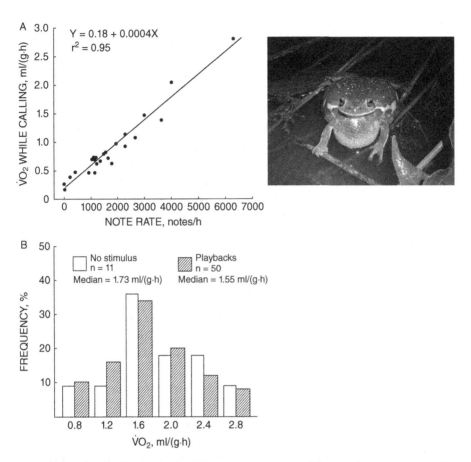

Fig. 1.1 *A*, A plot showing the relationship between note rate and the rate of oxygen consumption for the frog *Hyla microcephala* during vocalization. *B*, A frequency distribution of different oxygen consumption rates (VO_2) for the frog *Hyla microcephala* during vocalization. The open and filled bars indicate vocalizations in response to playback or without a stimulus, respectively. Note that, in *A*, the higher the note rate, the higher the rate of oxygen consumption. Redrawn from Wells and Taigen (1989) with permission from Springer publishing. Image is from Wikimedia Commons

Now let's examine a common predator-prey scenario that reveals both the importance and subtlety of animal performance for survival. Small fish often elude predators (e.g., a larger fish) by using rapid bursts of acceleration (Webb 1976), often in the stereotyped form of a C-start, a circular escape motion in which the fish swims off in the opposite direction from which it was first stimulated (by, for example, a touch on the tail, or the sensation of water pushing up against its tail). The salient point is that predators only have a very brief period of time to capture their prey, given this mode of rapid escape, and this time is usually on the order of milliseconds (a millisecond is one thousandth of a second). Whether the predator successfully captures the prey or, from the prey's point of view, whether the prey lives to see another day depends on many factors, including the initial attack distance and accuracy of the predator, the relative sizes of the predator and prey, and the relative performance capacities of each (i.e., how fast each can accelerate). Clearly, both predator and prey in such circumstances would evolve many specializations for maximizing their chances of success (capturing prey or eluding capture, respectively), but one of the most prominent specializations is the ability to accelerate quickly either to capture an evasive prey or to evade a quick predator. In other words, we can think of this simple yet ubiquitous dynamic as a coevolutionary arms race between contestants for performance supremacy. Performance often comes at a cost, so it isn't surprising to see this high level of performance relaxed when competition is absent. Indeed, if you find a lake where fish live without looming predators, their ability to accelerate could be reduced, as should the size of the anatomical parts that contribute to acceleration.

1.2 Definitions of performance

Before proceeding further, let's revisit the definition of performance capacity and flesh out some of the underlying assumptions. The ability to *quantify* a task is essential and distinguishes performance from other characteristics: one can measure how hard a dog can bite, count the number of seconds a frog will croak, measure the adhesive ability of geckos to cling to a surface, and quantify how long it takes Usain Bolt to run 100 m. This ability to quantify performance is essential for comparing different individuals, or comparing the same individuals at different times. However, based on this simple definition, many other kinds of performance can be defined, including how quickly an animal can digest food, how many offspring a female mammal can raise in a year, and so forth. In other words, "performance," if not carefully defined, can be extrapolated to a myriad of seemingly disconnected actions. From an evolutionary perspective, whole-organism performance holds little meaning if we attempt to compare "apples and oranges," such as comparing the maximum sprinting speed of a cheetah to the maximum book-reading performance of a teenage boy. In the former case, the task is essential for survival and pushes the entire physiological and muscular system to its limits whereas, in the latter case, extreme book-reading performance may be simply a function of a few heightened senses (e.g., eye coordination) and can hardly be considered as essential to survival (although we heartily advocate reading).

Because the goal of this book is to examine comparable kinds of performance that have significance in the lives of animals, we adopt a dynamic, functional, and whole-organism view of animal performance; such a view has been the center of conceptually related studies over the past several decades. The utility of this seemingly narrow definition can be understood by considering our own human Olympics, which measures human performance at a myriad of tasks, including sprinting, weightlifting, rowing, and so forth. However, the human Olympics do not measure many other potentially valid measures of performance, such as how many books an individual can read in a year or how much money a person can earn in a year. Similarly, this above definition of performance precludes suborganismal measures of performance, such as biochemical and physiological functions within organisms; these measures might include aspects such as the ability of enzymes to catalyze reactions, for example. A dynamic view of performance necessarily emphasizes movement, such as vocalization, locomotion, feeding. By contrast, aspects of performance such as digestion, metabolism, and other such measures are not considered here. In addition, we as humans intuitively consider performance to be a factor that can be increased with training, which would typically preclude things like digestion.

The logic behind this emphasis on the whole organism relates to the hierarchical nature of biological systems. Animals exhibit behaviors and functional capacities that are "emergent" properties at the organism level and cannot be fully understood by only examining individual components (fig. 1.2). Moreover, as noted above, it is the whole organism that is visible to the environmental forces that dictate life or death. Therefore, although understanding how an individual enzyme's function may provide some insight into a larger physiological process, it is the functioning of the larger system (e.g., the organism) that is of paramount importance for evolution. A final criterion is best illustrated by the Castanza family from the TV comedy "Seinfeld." Instead of celebrating Christmas, they celebrate "Festivus" in which "feats of strength" take place (alongside the presence of a bare iron rod). In other words, an important aspect

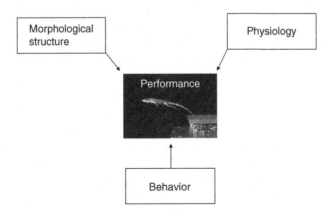

Fig. 1.2 A heuristic diagram showing the interrelationships between morphological structures, physiology, and behavior for influencing performance in animals. The image of the jumping lizard *Anolis valencienni* was taken by Esteban Toro with permission.

of this adopted definition is that high performance is both physically difficult and potentially costly. Sprinting at peak speeds is difficult, because of the
extreme coordination of numerous body parts (e.g., heart, lungs), and costly,
because of the energetic demands on the physiological system. The study of
performance would become meaningless if we were to compare performance
traits that are physically easy to those that are physically challenging.

The distinction between performance and nonperformance traits as defined
here can often be subtle based on the criterion of physical difficulty or cost. For
example, at first glance, there would seem to be little distinction between bird
song and frog vocalizations, but a closer look reveals some important differences. For many frogs, vocalization is energetically expensive, and long periods of vocalization are highly exhausting (Wells and Taigen 1989; Chappell
et al. 1995). Therefore, many frogs have evolved calling structures that allow
them to vocalize for long periods, with large amounts of slow-twitch muscle.
Because frog vocalization is challenging, it is also a basis by which female
frogs choose mates, with females selecting males that vocalize the longest.
In other words, frog vocalization is a "feat of strength" and, like many other
performance traits, varies among individuals. By contrast, bird song, while
often remarkably complex, is apparently not energetically expensive, at least
in some species (Chappell et al. 1995; Horn et al. 1995), although there is possibly variation among species in this regard. While there are anatomical and
structural costs to the evolution of bird song, many kinds of bird song can be
more analogized to piano playing than to an Olympian performance feat such
as running or jumping, although the concept of bird song is still useful for
demonstrating trade-offs, which we explore in chapter 7.

However, even if one accepts this definition, within the field of functional
morphology, one is confronted with an array of seemingly overlapping and
confusing terms. One distinction is between "performance" and "function." In
its broadest sense, "function" refers to the biological role of a morphological
trait (see Lauder 1996 for a discussion of this and other terms below). By biological "role," we mean the action that natural selection originally favored. Accordingly, "function" tends to be broadly defined, such as "grasping" or "manipulation" in the case of the human hand. This broad definition is essential
because morphological traits are often "multifunctional" and appear suited
for many different biological tasks. Consequently, a result of this definition
is that a single morphological trait can have multiple "functions," depending
on how specifically that function is defined. By contrast, performance is any
quantitative measure of *how well* a function is accomplished and thus provides
a necessary quantitative face on the rather amorphous concept of function.
For example, a key function of the lower limbs in bipeds is locomotion (e.g.,
jumping or running). We can *quantify* locomotion in any number of ways, such
as by measuring speed (how fast an animal runs), endurance (how long an

animal runs), and so forth, and therefore no single measure of performance will completely encompass all aspects of a broader function. Since dynamic movements arise from morphological structures, it is also useful to understand the relationship among function, performance, and two other terms describing morphological shape, namely, "structure" and "design." "Structure" refers to the underlying *arrangement* of parts in a particular morphological trait. By arrangement, we mean the configuration of parts, such as the attachment of tendons to bones, the arrangement of muscle fibers, and so forth. Throughout this book, we will refer to structure as the primary arrangement of morphological parts within organisms. By comparison, "design" is more difficult to define. Design overlaps with structure in referring to the arrangement of morphological parts, but the former refers also to an underlying (or predetermined) process, either conscious or unconscious. Because of this unfortunate historical connotation, we will no longer refer to design in this book. A final distinction is between performance and behavior. Behavior refers to what animals do, not how well they do it. An example from the human Olympics is instructive. In the high jump, Olympic jumpers used to jump over the bar belly first, whereas Dick Fosbury pioneered the "back-first" "Fosbury flop" that is widely used today (fig. 1.3). Hence, whereas the "behavior" of jumping has changed, the underlying metric of performance (jump height) has not. Note also the difference here between behavior and function, as jumping is a function, whereas how that jumping is done (belly first or back first) represents behavior, and how well that jumping is done (jump height) is the relevant performance trait.

Fig. 1.3 An image of the Fosbury flop. Image from Wikimedia Commons.

1.3 The hierarchical nature of biological systems

A pervading theme throughout this book is the concept of the hierarchy of biological systems; this hierarchy is manifested at two levels. First, performance capacity represents the final output from the underlying working components of complex functional systems. For example, many toads exhibit remarkable tongue-projection performance capacities, with accelerations approaching 300 m/s^2. Examination of the ecology of these toads provides insight into why these remarkable performance capacities have evolved. Many toads are largely defenseless because of their poor locomotor abilities and hence largely rely on crypticity (as well as some excreted toxins) to protect themselves from predators. However, this impaired locomotor ability also places the toads at a distinct disadvantage for capturing active prey, such as insects. Most toads have resolved this conflict by modifying evolutionarily a complex set of muscles and nerves that enable their tongues to project rapidly and accurately (Nishikawa 1999). However, this ability to project the tongue at extreme accelerations is only the final output derived from a complex set of interacting morphological structures. For example, key nerves innervate muscle groups that play a key role in tongue projection. The visual system in toads is also highly attuned to moving prey and thus is synchronized to track moving prey and enable accurate tongue projection. Thus, one cannot easily separate the underlying structures that produce performance from the evolution of performance capacity. Hence, while not dwelling on the precise mechanism of how performance is produced, we will examine the morphological and physiological structures associated with performance while also considering broader evolutionary and ecological implications.

The complex nature of morphological structures, physiological processes, and their effects on performance means that predicting how different morphological structures should influence performance is tricky. Because they emerge from a complex set of lower-level traits, performance traits do not always show proportional changes in relation to changes in traits such as morphology, behavior, or physiology. Small changes in morphology can have large effects on whole-organism performance, or vice versa. Such unpredictable effects are termed "nonlinear" because they don't occur predictably from one another in the way that one might expect in a classic linear relationship (Emerson et al. 1990). Two key factors that drive these nonlinear relationships are behavior and the environmental context of animal performance. For example, simple alterations in the way that animals use their body can have profound consequences on performance, even when the underlying morphology remains the same. This idea is familiar to anyone who has participated in organized sports—after all, how much time and effort is spent training athletes to move their bodies in certain ways to maximize success at a certain event? In the animal kingdom, where even slight differences in performance can make a difference between life and death, animals may make adaptive and conscious behavioral decisions that alter the basic relationship between morphology and performance. A classic example is flying frogs, which glide through the forest canopies of tropical forests in South America and Southeast Asia by the use of flaps of skin on their feet and flanks. Flying frogs cannot engage in powered flight like birds, but they can successfully glide for reasonably long distances by manipulating the position of their limbs, a discovery by Sharon Emerson and colleagues, who used plasticine models in wind tunnels to study the relationship between gliding ability and flap area (Emerson et al. 1990). They found that even slight alterations in the physical configuration of the models resulted in dramatic differences in gliding ability; but the effects of these changes were nonlinear, as other larger changes seemed to have less impact.

Second, measures of performance exhibit a hierarchical structure. For example, consider the case of crabs, as some males will fight, sometimes fiercely, to acquire and defend territories. Crabs exhibit several stages in fighting, beginning with an assessment stage, in which the crabs first evaluate the size and strength of their opponent, and then followed by intense grappling and wrestling with their two large front claws, which can result in severe injuries. In this system, one can define many potentially overlapping and hierarchical measures of performance. One could define performance relatively narrowly in measuring the pinching strength of claws by using force transducers, for example. Alternatively, one could take a broader view and measure fighting ability, which can be quantified as fighting success. However, note that, in this case, the broadest measure of performance (fighting success) is itself dictated by many other variables, including body size, the relative (i.e., size-adjusted) pinching strength of the claws, and so forth.

The choice of performance variables is as varied as the study subjects examined. Some researchers are interested in sexual selection and factors that determine male fighting success, or access to females, whereas other researchers are interested in understanding the mechanistic basis of how frogs jump. We will be inclusive in examining multiple, potentially disparate measures of performance capacity, for the reason that each measure can provide information on the underlying ecology and evolution of species. In some cases, we will even extract performance from studies that may not even have been considered their measures as performance. Below we describe several key examples that highlight this approach.

1.4 Variability, repeatability, and heritability

Three characteristics are vital for performance traits to play an important role during evolution: variability, repeatability, and heritability. Variability is the concept that a performance measure should vary among individuals within an interbreeding population. Performance capacities are distinctive from behavioral or morphological traits in sometimes (Le Galliard et al. 2004) displaying highly skewed distributions (fig. 1.4A), although many performance traits also show more normal-shaped patterns. This skewed pattern is generated by the presence of a large number of individuals who exhibit low or intermediate performance, and a few individuals who are outstanding performers. Contrast this pattern to the typical pattern for a morphological trait such

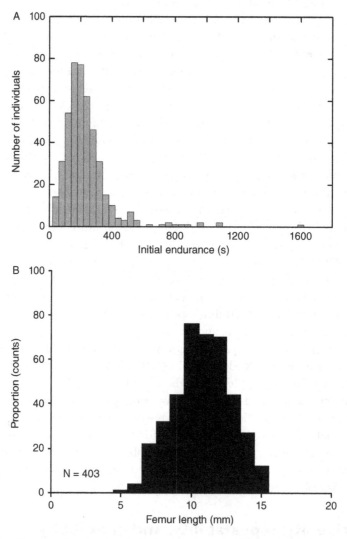

Fig. 1.4 *A*, A histogram showing the distribution of values of endurance for the lizard *Lacerta vivipara*. Taken from Le Galliard et al. (2004) with permission from Nature Publishing group. *B*, A histogram of femur lengths of green anole lizards, *Anolis carolinensis*, from Southern Louisiana (Irschick, unpublished data).

as limb length, in which the distribution is nearly bell shaped (i.e., normal; see fig. 1.4*B*). A second important feature of performance traits is repeatability, or the tendency for the same individual to consistently display the same level of performance capacity. Consider an example of two track stars running the 100 m dash, for which a world-class time would be 10 s. Further imagine that one of these two is an exceptional runner (track star A), with a personal best time of 9.90 s, and the other one is less exceptional (track star B), with a personal best of 10.20 s. If these fictitious track stars ran side-by-side 100 m dashes once per day for 5 straight days, and the times (in seconds) for runners A and B were (in order of days) 9.90, 9.89, 9.91, 9.90, and 9.88 (track star A), and 10.20, 10.22, 10.21, 10.20, and 10.19 (track star B), then it's easy to see (and statistical tests would verify this, but that's not necessary for this example) that maximum speed is repeatable in these individuals, with A being consistently better than B, no matter how often they race. Repeatability is an essential ingredient for performance traits to evolve because it means that a particular performance value for an individual is not a chance result but rather reflects basic physiological, morphological, behavioral, or genetic qualities.

A third factor is heritability, or the degree to which traits within a population have a genetic basis. A trait with a high heritability, typically measured on a scale of 0 (no genetic basis) to 1 (completely genetically controlled), is a trait that is likely to be inherited by an offspring from its parents. For example, consider a male alligator that has a particularly strong bite and happens to mate with a female who also has a very strong bite. If bite force shows high heritability (i.e., values closer to 1 than to 0), then the offspring of these two alligators will also likely be a hard biter, although other factors, such as the environment in which the offspring was raised, are relevant. All three of these traits (variability, repeatability, and heritability) are crucial to the evolutionary process, as traits that show high values for all three factors are also more likely to evolve via natural selection. By contrast, the efficacy of natural selection on performance traits that show low values for any of these three factors will be severely dampened.

1.5 The ecological context of performance

At the most basic level, performance capacities have evolved to enable animals to survive in their chosen habitats, whether in water, land, air, or the canopy of trees. Because of this basic requirement, body shape, physiology, and performance in animals are often matched to enhance chances of survival in different habitats. This phenomenon is glaringly obvious when one examines animals that occur in "extreme" environments, such as regions of extreme cold, heat, or salinity. For example, cave-dwelling organisms frequently possess enhanced organs and structures for touch or smell but reduced visual organs, whereas diurnal active hunters frequently rely on their visual abilities.

Large predators in the open landscapes of Africa or Australia often have excellent locomotor abilities (aerobic or anaerobic) that enable them to either ambush or chase down prey. Consequently, one can study the general nature of adaptation, or the process by which species evolve to match their habitats, by examining the performance capacities of species that occur in different habitats. Further, we can look within species and examine how individual variation in performance allows some animals to occupy some preferred habitats, while poor performers are more limited.

Let's ponder an example with butterflies, as it demonstrates nicely the ecological role of performance. A long-standing question in the evolution of flight concerns the effect of wing size on flying performance. Among others, two simple aspects of wing shape contribute to flight: the aspect ratio (wing length divided by wing width) and the relative loading of a system (body mass relative to wing area). Butterflies offer the opportunity to manipulate these parameters because one can surgically and nontraumatically remove certain amounts of the wing, as well as add loads, each of which can impair (though reversibly so) flight. Based on biomechanical predictions, one would expect that butterflies that have their wing area reduced should have reduced survival relative to butterflies whose wing areas were unmanipulated (Kingsolver 1999). In fact, laboratory studies seem to confirm this prediction; butterflies whose wing areas were reduced suffer dramatically reduced flight performance. But what are the consequences of this reduced flight performance for animals in nature? By releasing both manipulated (i.e., wings cut) and unmanipulated butterflies, Joel Kingsolver from the University of North Carolina made a surprising discovery; butterflies whose wings were surgically reduced did not suffer higher mortality rates compared to unmanipulated butterflies, despite the obvious performance disadvantage (fig. 1.5). Even butterflies whose wings were reduced by as much as 30% did not suffer reduced fitness relative to unmanipulated butterflies. This example highlights an important feature of the evolutionary process, namely, that evolution does not necessarily concern itself with decrements in performance if the animal can perform "well enough" in its environment to survive. In this case, it's clear that the butterflies, even with a large chunk of their wings missing, were able to move around effectively.

However, field observational studies of butterflies showed another competing role of butterfly wings, namely, thermoregulation. Many butterflies use their wings to thermoregulate; during cool mornings, butterflies spread their wings to gain maximum exposure to the rising sun and, during the hotter parts of the day, butterflies close their wings to avoid overheating. The reason for this careful thermoregulation is that flight is largely dictated by temperature. Flight performance is maximized across a relatively narrow range of temperatures, ranging from about 28°C to 32°C. Larger wings enable butterflies to effectively thermoregulate by increasing the surface area that is exposed to the

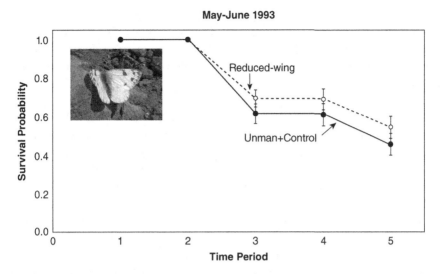

Fig. 1.5 Survival probabilities for the butterfly *Pontia occidentalis* (*inset*), for animals whose wings had been reduced in size (dashed line) and for those butterflies whose wings had not been altered (solid line). Note that despite the small difference in survival probability, there was no significant difference between the reduced-wing and control animals. Redrawn from Kingsolver (1999) with permission from Wiley-Blackwell Press. Image of the butterfly is from Wikimedia Commons.

sun. Therefore, an ecological perspective on performance (in this case, flight) can reveal the complex nature of morphological traits, and the performance traits they have evolved for. A priori, one would expect that the most obvious function of butterfly wings would be flight, but the presence of other roles (e.g., thermoregulation) suggests that not all aspects of wing morphology have evolved as adaptations for flight. This plurality of morphological structures is a recurring theme in the evolution of performance.

1.6 The evolution of performance

One of the most notable features of the animal kingdom is the tremendous variability within and among species in body shape and performance. For example, within anurans (frogs), there is variation in the length of the hindlimb (which directly powers jumping in frogs) and in jumping ability (i.e., how far a frog can jump). Many frogs can only jump a few inches, but the record frog jump at the Calaveras Frog Jumping Contest was over 2 m long (Astley et al. 2013)! Whereas we can learn a great deal about the muscular and neural factors involved in jumping by studying single species, we can learn far more about how jumping has evolved by taking a comparative approach. The comparative approach in evolutionary biology is a heuristic way of testing ideas about adaptation by comparing species in divergent environments and thereby testing how different selective regimes have resulted in different morphological and performance traits. For instance, when one compares species of fish that occur in polar regions with those from more temperate areas, one of the reasons why the polar fish can survive in freezing waters (without freezing) becomes obvious: some arctic fish possess ice-nucleating agents that prevent their internal tissues from freezing (and therefore killing the fish), whereas fish in warmer regions typically don't require such an adaptation. Only by comparing different species can one clearly understand the connection between the environment and adaptive features for avoiding freezing. Just as we can use the comparative approach to understand why some fish can survive freezing and others cannot, we can profitably apply this idea toward explaining variation in performance capacity among and within animal species.

One of the most useful ways to study how traits evolve is to understand the selective pressures that have driven the change in the first place. One of the driving forces for variation in locomotor performance is the threat of predation. Because many animals flee predators using locomotion, much of the variation in peak running speed has evolved as a direct response to pressure from predators of varying locomotor abilities and strategies. One instructive example comes from studies of escape strategies in damselflies, genus *Enallagma*, that were examined in the Northern Lakes region in the USA by Mark McPeek and his colleagues (McPeek et al. 1996; McPeek 1999). There are two important predators for these invertebrates: dragonfly larvae and fish, each of which requires different strategies for effective escape. The ancestral and predominant ecological context for damselflies is the occupation of lakes with fish predators but, in some cases, damselflies have successfully colonized lakes that contain a novel predator (dragonfly larvae). Whereas damselflies avoid active locomotion as an escape strategy in the presence of much faster and larger fish (doing so would only provide a quick snack!), rapid bursts of locomotion are more effective as a strategy for eluding the ambushing dragonfly larvae. This novel and intense form of predation pressure from dragonfly larvae has had several physiological, morphological, and functional evolutionary consequences for damselflies, including the evolution of a greater number of tail fins for propelling locomotion (called lamellae) as well as the evolution of enhanced burst speeds, and enhanced biochemical activity for several key enzymes that enhance burst speed.

Many aspects of morphology and physiology influence burst speed; however, in this case, three enzymes seem to play an especially important role: pyruvate kinase, lactate dehydrogenase (both used in glycolysis), and arginine kinase (involved in reestablishing the pool of available ATP). Fitting a priori predictions, damselflies that have colonized lakes with dragonfly larvae predators show significantly greater activity for arginine kinase, but not for the two other enzymes. Because these damselflies move only in brief bursts when escaping, the increased activity of arginine kinase may extend the period of maximal exertion for several seconds, providing a crucial window of opportunity. Although indirect, this example suggests that variation among species in arginine kinase activity (physiology) is adaptive because of its role in influencing burst speed (performance) that has arisen as a consequence of variation of different habitats (lakes with fish predators vs. lakes with dragonfly predators). Further, this damselfly example demonstrates the coevolutionary interaction between prey and predators and shows how the presence of different predators can alter escape behavior (rapid escape vs. lying still) and consequently both performance levels (high vs. low burst speed) and physiological characteristics.

1.7 Behavior and performance

As exemplified by the Fosbury flop, described in section 1.2, behavior and performance are often inextricably linked. Our thesis is that behavior plays a key role in determining both whether animals *undertake* a particular performance task and how *effective* an animal is at a particular performance task. Particularly germane to this idea is the notion that animals make conscious choices about how to move their body to maximize certain aspects of performance (and perhaps minimize other, more deleterious aspects). The idea of behavioral choices, or cognition, in animal performance is a relatively new one. The field of biomechanics, the study of the physics of animal movement, stresses the stereotyped nature of neuromuscular movements and examines the optimal ways in which animals can achieve peak performance by moving in a specific manner. However, functional morphologists are now aware that individuals employ different strategies for maximizing performance and that there may not always be "one" way to perform a particular task. This phenomenon becomes obvious when comparing different individuals or species with different phenotypes. A popular example comes from human athletics: the relatively slight Janet Evans found success during swimming events by rapidly cycling her arms like a windmill in order to keep up with (and defeat in many cases) much larger swimmers who took fewer, longer arm and leg strokes to swim. In her case, the decision to adopt a different swimming strategy was a conscious and successful modification of behavior. The same is true for professional basketball, which has undergone an evolution with respect to the way in which free-throw shots are performed. Rick Barry, a member of the NBA Hall of Fame, is famous for employing an underhand shooting method and achieving a 90% success rate throughout his career. The fact that his approach has now been completely replaced by the overhand shot, with great success, again highlights that there can be multiple ways through which to achieve the same level of performance.

One performance trait thought to be stereotyped is jumping; but even for this task, there is evidence for the overarching role of behavior. Jumping has been particularly well studied in frogs and lizards, both of which commonly use jumping to escape predators or simply move around their habitats. One group of lizards in which jumping has been well studied is the diverse genus *Anolis* (Irschick et al. 1997; Losos 2011). The arboreal insectivorous lizards from this genus present a wide diversity of morphological forms. Some species have long hindlimbs and are fast, active runners, whereas other species have short hindlimbs and are slow, cryptic twig specialists (fig. 1.6). Such morphological and behavioral adaptations have occurred largely because the different species occupy arboreal perches of different diameters and heights. Whereas some species occur on broad perches (e.g., tree trunks) close to the ground, other species occur high in the forest canopy on narrow twigs. These different surfaces impose divergent functional demands for movements such as running and jumping. Jumping is an important activity for these arboreal animals, as up to 50% of their movements will be jumps, often between branches. However, anole species vary in how far they can jump, with some species being outstanding, and others mediocre. Why do anole species vary in jumping capacity, and which factors enable some species to be better jumpers than others?

A basic understanding of the biomechanics of jumping is necessary to answer this question. Jumping in limbed vertebrates can be deconstructed into several stages (fig. 1.7). Four primary stages exist: the preparatory phase, the takeoff phase, the suspended phase, and the landing phase. How far a lizard can jump is largely a consequence of three factors: the takeoff velocity, the

Fig. 1.6 Images of anole lizards, *Anolis* spp., from the Caribbean. Note the diversity in body form and habitat use, including variation among species in head, body, limb, and tail dimensions. The titles above each lizard indicate their preferred habitat. Images by Duncan J Irschick.

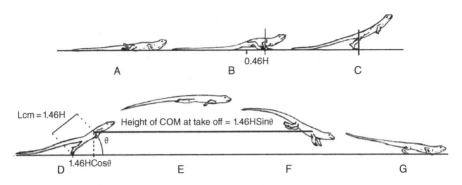

Fig. 1.7 A diagram showing the typical profile of jumping for a limbed vertebrate, in this case a cartoon lizard. Note that the key variables for increasing jump distance are takeoff angle, hindlimb length, and takeoff velocity. Taken from Toro et al. (2004) with permission from the University of Chicago press.

hindlimb length of the animal (as jumping in lizards occurs largely via the hindlimbs), and the angle of takeoff. A longer hindlimb enables a lizard to accelerate its mass over a greater distance and hence increase its overall horizontal jump distance. Similarly, a higher takeoff velocity also should increase jump distance because force is a function of mass and acceleration ($\mathbf{F} = m\mathbf{a}$), and a higher velocity typically translates into a higher acceleration. Indeed, for a lizard of a known hindlimb length and takeoff velocity, one can predict the "optimal" takeoff angle that a lizard should use to jump the maximum theoretical distance.

One can test these ideas by examining jumping across a variety of *Anolis* lizard species. One way of examining jumping is by inducing lizards to jump off a force platform, which measures the forces that an animal exerts as it jumps. Studies of a group of 12 anole species (Toro et al. 2004) have shown that lizards do not typically jump at the "optimal" angles for maximizing horizontal jump distances (fig. 1.8). Rather, most anole species jump, on average, several degrees lower than expected. Why would anoles jump at such "suboptimal" angles? Part of the answer may reside in the relative benefits and costs associated with jumping at different angles. By jumping only a few degrees lower than predicted, anoles suffer only a very minor decrement in horizontal distance (average loss among species = 1.4%), but they dramatically diminish the vertical height of their jumps (average loss = 15.1%), as well as the duration (average loss = 7.3%). In other words, by jumping only a few degrees less than their predicted "optimal" angles, anoles can achieve relatively long (in distance), short-lived, and shallow jumps.

Why would anoles want to jump with short durations and low jump heights? Understanding the natural habitat of these lizards is crucial for interpreting this result. Arboreal *Anolis* lizards occupy dense matrices of branches

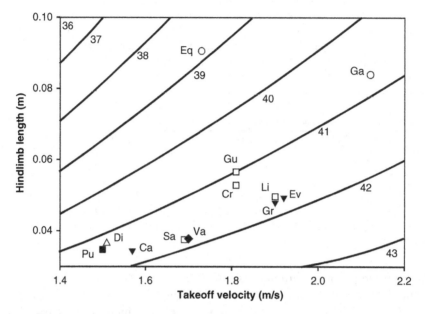

Fig. 1.8 The relationship between hindlimb length and takeoff velocity for 12 anole lizard species. The lines show the "optimal" takeoff angles for species for combinations of hindlimb length and takeoff velocity. The dots show where species reside. Abbreviations refer to different species. Note that very few species reside on the optimal lines and that actual takeoff angles typically fall slightly below the predicted optimal angles. Taken from Toro et al. (2004) with permission from the University of Chicago press.

and vegetation that must be negotiated effectively and, by employing much shallower and shorter jumps, anoles reduce both the likelihood that they will hit vegetation during jumping, and the likelihood that they will be captured in midair by predators. Further, they can accomplish this by suffering only a very minor decrement in jumping distance. The salient lesson from this example is twofold. First, one cannot easily separate the evolution of behavior from the evolution of performance. While morphological features may limit or enhance performance, behavior is equally important for defining what animals actually choose to do, as opposed to what they might do based on theoretical considerations. A longer hindlimb enables lizards to jump farther, but these lizards also made behavioral choices about how they wanted to jump that influenced many parameters, such as the angle of takeoff. Second, to interpret differences among species in performance, it is useful to consider the ecological conditions in which that performance trait has evolved. For *Anolis* lizards, the dense matrix of branches forms a very different jumping landscape compared to an open landscape where only jumping long distances would be important. In such a dense matrix, the most important metric of performance is likely not distance per se but the ability to jump short distances accurately and without running into other branches.

1.8 Mechanistic and energetic constraints on performance

Because they emerge from underlying structural and physiological features, there are often severe structural and neuromuscular constraints on performance capacities. For example, for animals that rely on muscles to move, the physiology of the muscles limits both the speed and amplitude of movement, at least most of the time (see chapters 4 and 5). Because muscle fibers are activated by neurons that innervate them, the rate of nervous transmission is one of the primary drivers of muscle function. In addition, because any kind of dynamic movement requires energy, knowing the ability of organisms to obtain and process energy is crucial to understanding limits on animal performance. One of the ways to study the role of the neuromuscular system in animal performance tasks is to examine "extreme" performance, in which animals accomplish performance feats that are seemingly beyond their inherent capacities. By doing so, one can often gain insight into not only the neuromuscular constraints on animal performance but also how those constraints can be broken.

Consider the role of anatomical structure and the neuromuscular system on tongue projection in frogs and salamanders. Unlike many animals that rely on rapid locomotion to capture prey, many salamanders and toads rely on rapid projections of their tongues to secure mobile prey. How rapid? Some frog species can project their tongues in time periods as short as 30 ms, or 30 one-thousandths of a second. Some salamander species can project their tongue at high velocities and over large lengths (up to 100% of body length) by using a unique anatomical arrangement that ballistically propels the tongue forward (fig. 1.9; Deban et al. 1997; Deban and Dicke 1999). A perhaps even more amazing example comes from chameleons, whose tongues also exhibit ballistic properties and can extend up to twice their body length. For both of these groups, these dramatic tongue-projection abilities allow them to overcome their own sedentary nature to capture a wide range of active prey.

The unique structure and function of extreme tongue protrusion shows how extreme performance often requires a radically novel readjustment of available morphology, thereby allowing animals to "break the rules" that "normal" physiology would typically allow. This theme of morphological novelties enabling animals to achieve unusually high levels of performance is a recurring one in the animal kingdom, and one that will be revisited in chapter 9 ("Extreme performance"). This book will also explore the more basic roles of anatomical structure and the neuromuscular system in driving the evolution of animal performance.

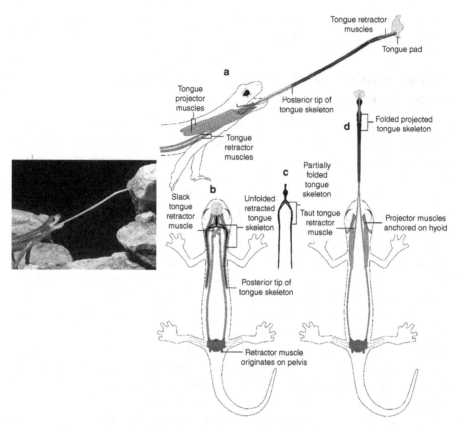

Fig. 1.9 *Left*, Image of a *Hydromantes* salamander. *Right* (*a–d*), The internal anatomy of the tongue-projection system in a *Hydromantes* salamander, showing how the tongue can be projected at high velocity and length via ballistic principles and then retracted back into the mouth. The mechanism of tongue projection is similar to that of squeezing a wet watermelon seed between one's fingers. Taken from Deban et al (1997) with permission from Nature Publishing group.

References

Astley HC, Abbott EM, Azizi E, Marsh RI, Roberts TJ. 2013. Chasing maximal performance: A cautionary tale from the celebrated jumping frogs of Calaveras County. Journal of Experimental Biology 216:3947–3953.

Chappell MA, Zuk M, Kwan TH, Johnsen TS. 1995. Energy cost of an avian vocal display: Crowing in red junglefowl. Animal Behaviour 49:255–257.

Deban SM, Wake DB, Roth G. 1997. Salamander with a ballistic tongue. Nature 389:27–28.

Deban, SM, Dicke, U. 1999. Motor control of tongue movement during prey capture in plethodontid salamanders. Journal of Experimental Biology 202:3699–3714.

Emerson SB, Travis J, Koehl MAR. 1990. Functional complexes and additivity in performance: A test case with "flying" frogs. Evolution 44:2153–2157.

Gleeson, TT. 1979. The effects of training and captivity on the metabolic capacity of the lizard *Sceloporus occidentalis*. Journal of Comparative Physiology 129:123–128.

Horn, AG, Leonard, ML, Weary, DM. 1995. Oxygen consumption during crowing by roosters: Talk is cheap. Animal Behaviour 50:1171–1175.

Irschick DJ, Vitt LJ, Zani P, Losos JB. 1997. A comparison of evolutionary radiations in Mainland and West Indian *Anolis* lizards. Ecology 78:2191–2203.

Kingsolver JG. 1999. Experimental analyses of wing size, flight and survival in the western white butterfly. Evolution 53:1479–1490.

Lauder GV. 1996. The argument from design. In Rose MR and Lauder GV, eds. Adaptation. Academic Press, San Diego, CA, pp. 55–91.

Le Galliard JF, Ferrière R, Clobert J. 2004. Physical performance and Darwinian fitness in lizards. Nature 432:502–505.

Losos JB. 2011. Lizards in an Evolutionary Tree: Ecology and Adaptive Radiation of Anoles. University of California Press, Berkeley, CA.

McPeek MA. 1999. Biochemical evolution associated with antipredator adaptation in damselflies. Evolution 53:1835–1845.

McPeek MA, Schrot AK, Brown JM. 1996. Adaptation to predators in a new community: Swimming performance and predator avoidance in damselflies. Ecology 77:617–629.

Nishikawa KC. 1999. Neuromuscular control of prey capture in frogs. Philosophical Transactions of the Royal Society of London, Biological Sciences 354:941–954.

Toro E, Herrel A, Irschick DJ. 2004. The evolution of jumping performance in Caribbean *Anolis* lizards: Solutions to biomechanical trade-offs. American Naturalist 163:844–856.

Webb PW. 1976. The effect of size on the fast-start performance of rainbow trout *Salmo gairdneri*, and a consideration of piscivorous predator-prey interactions. Journal of Experimental Biology 65:157–177.

Wells KD, Taigen TL. 1989. Calling energetics of a neotropical treefrog, *Hyla microcephala*. Behavioral Ecology and Sociobiology 25:13–22.

2 The ecology of performance I
Studies of fitness

2.1 Natural selection, sexual selection, and performance

As mentioned in chapter 1, natural selection is the differential survival of individuals within a population in response to some external environmental force, such as temperature, predators, and so forth. As such, natural selection is a key regulator of animal and plant populations and is a driver of evolutionary change; therefore, it is responsible for much of the diversity we see today. Over the last hundred or so years, this concept of natural selection has been enormously valuable, both conceptually and quantitatively. Although the concept of natural selection was first elucidated by Charles Darwin and Alfred Russell Wallace, the importance of this process remained largely opaque until a seminal review by John Endler (1986) showed that natural selection in the wild was both abundant and strong. This paper paved the way for a whole new generation of studies that have been able to show how natural selection shapes animal populations in a myriad of ways. There are two primary ingredients for natural selection. First, there must be variation among individuals in traits that are important to survival, such as morphology, performance, and behavior. Second, more individuals are born into a population than can possibly survive, given limited resources. Natural selection represents the nonrandom process by which some individuals live and others die, with some aspect(s), such as performance, driving mortality. However, one must be careful to distinguish between selection and evolution. Selection consists of nonrandom survival across generations and makes no assumption regarding the genetic basis of the target/focal traits. By contrast, evolution consists of a

Animal Athletes. Duncan J. Irschick and Timothy E. Higham. © Duncan J. Irschick and Timothy E. Higham 2016. Published in 2016 by Oxford University Press.

change in the mean value of a trait across generations, and this change need not occur purely by selection but may also occur by other processes, such as random genetic drift (essentially nonadaptive random evolution), correlated responses due to pleiotropy (a single gene having multiple impacts on seemingly unrelated phenotypic traits), or linkage disequilibria (nonrandom association of alleles). Consequently, it is possible for a trait to be under intense selection without any evolution if the trait has no underlying genetic variation. However, most morphological, behavioral, and performance traits do have some genetic basis (see chapter 11) and, hence, there is good reason to expect that any kind of significant natural selection on such traits should result in an observable evolutionary change, depending on the strength of selection.

If one believes that performance traits are important to animals in their natural habitats, then one would predict that natural selection should favor individual animals with high performance capacities; however, in reality, performance traits may not evolve under such a single-minded principle. Unlike the human Olympics, in which the fastest or strongest individual is always the winner, performance traits in nature evolve within a more complex social and ecological environment that often demands trade-offs, which often keep performance traits from being optimized. Keep in mind that "peak" performance is ultimately a human construct, and one should be wary of imposing such views on animals (an exception is being chased by a bull, in which case your speed becomes very important!). In some cases, there is evidence that a particular performance trait is important to animals in nature because of an extensive knowledge of the natural history of the species. For example, we know that cheetahs rely on sprinting quickly to capture elusive prey, and therefore it seems logical to assume that a particularly fast cheetah will capture more prey and thereby be more likely to live longer, enjoy good health, and so forth, compared to a slow cheetah. However, whether high speeds enable other animals that may rely more on stealth than speed to survive is an open question. Studies of natural history are therefore important because they provide background information that enables researchers to form hypotheses about how natural selection operates on animal populations. The most direct method of assessing whether a performance trait is important in nature is to conduct field studies to determine whether unusually good or bad performance correlates with fitness.

2.2 Path diagrams of selection

A significant challenge for biologists who study selection is to understand the true "targets" of natural selection. Consider the above example of a cheetah running after a swift gazelle. While it may be obvious that "survival of the fittest" will favor the fastest cheetah, we would also like to know a few other things. Do cheetahs with long hindlimbs tend to capture more gazelles because long hindlimbs propel gazelles to higher speeds? Do cheetahs that only stalk in the early evening (and not the morning) capture more prey? The point is that many aspects of morphology, behavior, and performance are likely to influence whether a cheetah lives or dies, and how many of their offspring survive to the next generation. Further, many traits may be tightly correlated; perhaps cheetahs that have long hindlimbs (and therefore run fast) also tend to hunt during only particularly favorable times of the day, thereby increasing their likelihood of capturing prey. Additionally, the fast speeds of cheetahs appear to correlate strongly with maneuverability, another important predictor of predator-prey outcomes. It has been suggested that the fast speeds that cheetahs are capable of obtaining are simply a byproduct of natural selection favoring increased maneuverability. In fact, cheetahs don't typically run very fast when they're hunting in nature (Wilson et al. 2013), and the ability to turn easily at high speeds may make more of a difference to them for capturing prey than straight-line speed would. Consequently, one must be cautious of solely focusing on particular traits without considering the broader ecological context. Studies of natural selection are also useful for gaining insight into other aspects of animal ecology and evolution. For instance, one might ask, why do bats have large ears, or why do burrowing lizards have numerous vertebrae? If we can understand whether selection favors ear size or vertebral number within a species, we might be able to gain some insight into why some species have such traits and not others.

The problem of intercorrelation among traits presents a logistical and statistical challenge. Therefore, evolutionary biologists have designed statistical methods for disentangling the various ways in which selection operates on different kinds of traits (Lande and Arnold 1983; Schluter 1988). One way that researchers can study how selection acts is through the use of path diagrams, which are conceptual tools for understanding linkages among traits. Classic research by Steve Arnold (1983) allows us to first examine how variation in performance traits influences variation in fitness (fig. 2.1A). This basic formulation can be modified to create a more detailed heuristic path diagram. Such diagrams have been recast in several ways, but one valuable framework (Kaplan and Phillips 2006) is to examine how selection acts on five distinct aspects of the organism: the environment, development, morphology, performance, and fitness (fig. 2.1B). Developmental and environmental processes will interact to produce phenotypes upon which selection can act. In turn, variation in the phenotype has a strong effect on performance, which in turn directly affects fitness. Importantly, selection can act on various factors within this diagram; for instance, selection could eliminate defective phenotypes that die as a result of developmental abnormalities, or selection could eliminate animals

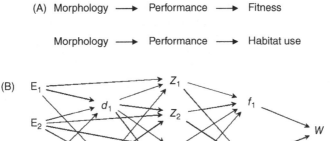

Fig. 2.1 *A*, Heuristic diagrams showing relationships between morphology, performance, and fitness. This paradigm allows researchers to examine variation among individuals in morphology and performance and relate them to variation in fitness (Arnold 1983). The second diagram is an elaboration of the one above it but is better suited for variation among species, with relationships between morphology, performance, and habitat use (Garland and Losos 1994). *B*, A holistic view of the factors that influence fitness variation among individual animals. Taken from Kaplan and Phillips (2006) with permission from Wiley-Blackwell Press.

that possess inferior adult phenotypes with subpar performance capacities. This framework also does not take into account selection on behavior, which can play a direct role in how selection operates. Consider again our hypothetical cheetah, and imagine an individual animal that always attacked its prey at night, when visibility and chances of prey capture were low. Such a behavior might rapidly result in any animal being quickly eliminated from the population simply because it could not capture sufficient food to survive.

2.3 Definitions of fitness

Any study of natural selection must also grapple with the notion of fitness, which is complex but, for the sake of the discussion here, is defined as the ability of an individual to produce offspring that then go onto producing more offspring, and so forth. In practical terms, few studies can meet this definition per se, and more indirect measures are typically examined. One measure of fitness is survival across a time period (e.g., a year), with the assumption that, if an individual survives across a year, they are more likely to pass on their genes to the next generation via mating. By contrast, an animal that dies, and that does not reproduce, has a net fitness of 0. A second common measure is paternity, which is usually assessed by genetic techniques that determine how many offspring an individual has sired (father) or given birth to (mother). In some cases, researchers can assess the lifetime paternity, though more often, researchers estimate this lifetime paternity by sampling across a time period in the life of the animal (e.g., from the breeding to the nonbreeding season). Other measures include growth rate, body condition, or any other measures believed to be related to the likelihood of passing on genes to future generations.

In practice, researchers estimate selection on traits by assessing the relationship between the fitness measure and the trait in question. Some laboratory studies of natural selection measure fitness in a slightly different way, namely, by measuring the rate of increase of the population (such as in bacteria); however, here we will focus on the former method. An example comes from Hermon Bumpus who in 1899 published a paper in which he described morphological traits in a group of birds before and after a severe storm and then determined which traits dictated whether a bird survived (Bumpus 1899). Other indirect ways of measuring fitness focus on aspects of reproductive success. For example, in territorial animals, the number of females that a male can mate with is one, indirect measure of reproductive success. However, such indirect measures present their own assumptions. For example, a male animal might not mate with, or produce viable offspring with, every female that occurs in his territory. In some mammal species, females store sperm, and sperm competition (literally, competition among sperm from different males that mate with the same female) can result in the sperm from some males being better able to fertilize females. In sum, there are many ways that researchers will attempt to estimate fitness, but all generally aim to understand the ability of parents to produce offspring that will then go onto producing more fecund offspring.

2.4 Quantifying the impact of natural and sexual selection

The most ubiquitous method for measuring selection is using mark-recapture studies on individuals in the field. Researchers initially mark a sample of individuals from a population and then resample the population at some later period (e.g., 6 months or a year later). Survivors are then compared against the nonsurvivors for some trait of interest (e.g., morphology, performance trait). One can also combine mark-recapture with genetic studies of paternity to establish another measure of fitness (reproductive success), and some studies have found that the likelihood of annual survival and the likelihood of reproductive success are not necessarily correlated in a 1:1 manner. The mark-recapture method offers the advantage that natural selection is quantified in a natural setting, which does not occur in studies with seminatural, or enclosed environments. Understanding how selection works on organismal traits in a more natural setting is crucial, especially since variation in animal mortality and reproductive success is often correlated with stochastic or seasonal variation in environmental conditions (e.g., light levels, rain levels). Some problems with the mark-recapture approach include the need for large sample sizes for robust statistical analyses, and the large commitment of time that must be spent to mark and recapture the bulk of the population. Statistical analysis of selection requires large sample sizes (typically >100 initially marked individuals) to effectively estimate the direction and strength of selection and, therefore, biologists often focus on large populations. However, successfully marking the majority of individuals in a large population is challenging, especially if there is dispersal into and out of the population. Therefore, in many cases, small closed populations in which emigration and immigration are limited (as on an island) offer the best opportunities for measuring natural selection.

Mark-recapture field studies are not the only way of studying the effects of performance variation on fitness. An interesting approach consists of semi-natural experimental arenas that represent "halfway houses" between controlled environments and natural habitats. Such experiments typically consist of large enclosed arenas in which a sample of animals can be easily marked and recaptured and into which predators can be introduced. Small aquatic organisms, such as fish and tadpoles, are fitting systems for this approach, as these animals often occur naturally in small, enclosed ponds. An advantage of this seminatural approach is the ability to manipulate and measure the target population at higher levels of detail, as compared to the case in field situations. In the field, scientists are usually able to mark only a subset of the population of interest, and continual immigration and emigration are important sources of error. By comparison, in seminatural experiments, scientists can mark and measure every individual in the experiment.

2.5 Does high performance result in higher fitness?

A widely held belief, among both biologists and the general public, is that superior performance should result in superior fitness. In other words, individuals within populations that are particularly good performers should be more likely to survive from one year to the next and, ultimately, should be rewarded with high reproductive success. The logic of this view seems intuitive; but it is also possible that, because of trade-offs with other traits, one could observe stabilizing selection on performance traits, that is, that natural selection would favor intermediate (average) performers and disfavor both extremely good and poor performers (fig. 2.2). By contrast, there is little reason to anticipate either disruptive or negative directional selection on performance traits, as this would suggest that at least some poor performers would be favored by natural selection. In surveys of various kinds of traits (mostly morphological traits, but some behavior and life-history traits, fig. 2.3), performed by Joel Kingsolver

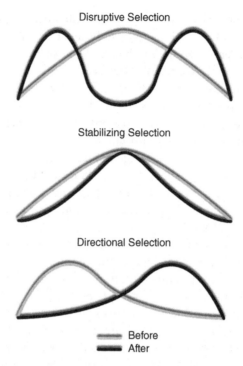

Disruptive Selection

Stabilizing Selection

Directional Selection

Before
After

Fig. 2.2 Depictions of three major kinds of natural selection, including (from top to bottom), disruptive, stabilizing, and directional selection. Note that the "before" distributions are one of many possibilities. The distributions refer to a hypothetical abundance of phenotypes before and after selection. Taken from Wikimedia Commons.

and his colleagues (2001), directional selection is the most common kind of selection in nature, whereas stabilizing and disruptive selection are rarer. While fewer studies of selection on performance have been carried out, the available data (Irschick et al. 2008) are concordant with these findings: in a survey of 23 studies, about half showed significant and positive (directional) selection on performance. In other words, selection often (but not always) favors the best performers and rarely favors mediocre or bad performers. An interesting case study of this comes from studies of juvenile collared lizards, *Crotaphytus collaris*, by Jerry Husak and colleagues (2006). Juveniles of many species are often under intense predation pressure, as they are often vulnerable compared to larger, faster, and stronger adults. At least in juveniles, rapid locomotion is often important for evading predators and, therefore, faster juveniles should enjoy a survival advantage, as compared to slower individuals. Juvenile collared lizards, *Crotaphytus collaris*, seem likely to suffer from this phenomenon, as large adults can pack a powerful bite, whereas smaller and weaker individuals are likely to be more vulnerable. Husak and his colleagues

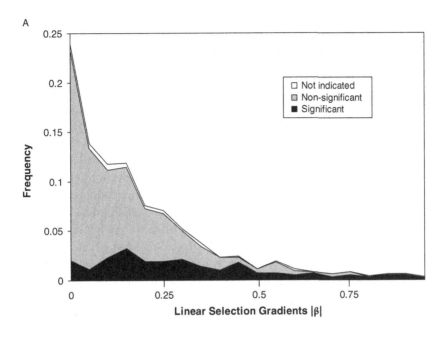

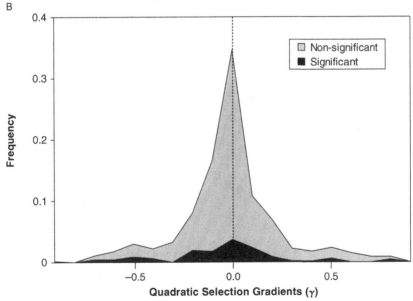

Fig. 2.3 *A*, A histogram of linear (directional) selection gradients for a wide variety of traits. This histogram depicts the relative strength of directional selection through both the magnitude and the number of significant versus nonsignificant values, as shown in different shading. *B*, A histogram of quadratic (stabilizing) selection gradients showing the relative strength of stabilizing selection. Taken from Kingsolver et al (2001) with permission from University of Chicago Press.

found that indeed there was strong selection favoring high sprint speeds in juvenile but not adult collared lizards.

What about other kinds of selection? Stabilizing selection implies a constraint or trade-off on performance that imposes a cost against individuals with high or low values of performance. One example is reproduction, in which animals produce eggs or offspring. The extra mass and physiological changes can alter how selection operates on performance. Some lizards produce large numbers of eggs, which can impair locomotor performance and thereby alter how selection operates on females with eggs. In side-blotched lizards, *Uta stansburiana*, this may explain why there is stabilizing selection on endurance capacity in females (Miles et al. 2000). Female lizards have reduced endurance capacity because of the burden of possessing eggs, which add substantially to both the mass and overall size of these small lizards. As further evidence, once follicles are experimentally removed before ovulation (thereby reducing the reproductive burden experienced by females), female lizards increased their endurance capacity (fig. 2.4). The experimental increase in endurance capacity relaxed some of the mortality costs arising from reproduction, except in the case of some high-performance females. Another example of possible stabilizing selection comes from a study done on garden snails, *Helis aspersa* (Artacho and Nespolo 2009). In that study, the authors of that study examined

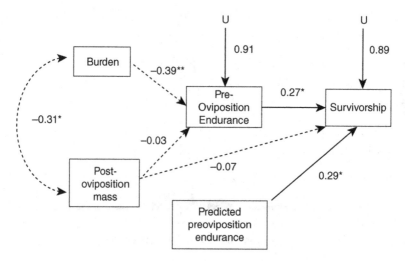

Fig. 2.4 Interactions among various life-history traits, including reproductive burden, and survivorship in side-blotch lizards, *Uta stansburiana*. Significant relationships are noted by thick lines and bolded numerical values. Numerical values refer to correlation coefficients. Note the strong positive relationship between predicted oviposition endurance and survivorship, and the strong negative relationship between reproductive burden and pre-oviposition endurance. From Miles et al. (2000) with permission from Wiley-Blackwell Press.

selection on standard metabolic rate, that is, the minimum physiological cost of maintenance, on snails in a seminatural enclosure. Among other results, the researchers found that average-to-poor metabolic rates were favored by natural selection in these snails, a result that suggests that snails that are too "slow" (low metabolic rate) or too "fast" (high metabolic rate) would be vulnerable to predators or would burn through energy too quickly, respectively.

Whether a performance trait influences survivorship or reproductive success can depend on how these traits affect tasks that might appear to have some conflict, such as eluding predators versus attracting females. In a recent series of studies, Jerry Husak and colleagues have investigated the factors responsible for influencing survival and reproductive success by carrying out mark-recapture studies in collared lizards, *Crotaphytus collaris*. These are robust and hard-biting lizards that use their powerful jaws to defend territories from rival males (as well as capture and consume prey) and use rapid locomotion both to capture prey and to elude predators. Husak and colleagues (Lappin and Husak 2005; Husak 2006; Husak et al. 2006) addressed the adaptive value of two performance variables (sprint speed and bite force) and several morphological variables (head shape and hindlimb length), as well as of the hormone testosterone. Because testosterone is known to build muscle in vertebrates, it could potentially explain why some collared lizards bite hard and run fast, whereas others are more wimpy. Testosterone can also increase behavioral aggression, which may enhance the ability of males to defend territories and therefore mates. They found that large adult male collared lizards with high bite forces and high sprint speeds had more offspring, relative to more wimpy performers (fig. 2.5), but these performance traits offered no advantage for survival (also, there was also no selection on testosterone levels). This study suggests that, in adult male lizards, being a good athlete means having increased access to females, probably because these males can effectively keep out rival males, and perhaps because females prefer high-performance males (though this has not been shown). However, being a good lizard athlete does not mean one is more likely to survive from year to year, suggesting that these traits in this species may be more important for sexual selection than natural selection.

In some cases, there is also evidence for limitations on how much high performance is favored. *Urosaurus* lizards are small, fleet, insectivorous lizards common in the deserts of the Southwestern United States. Mark-recapture data show that, as with adult male collared lizards, fast male lizards are more likely to survive from one year to the next if they are fast runners, but there is also a limit on the strength of this selection (Irschick and Meyers 2007). Close inspection of the fitness surface shows that selection on sprint speed shows both directional and stabilizing components, that is, that it's generally bad to be a poor runner, but it's also bad to be an *exceptionally* good runner. In other

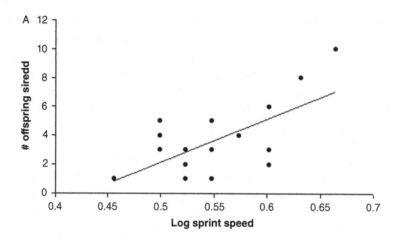

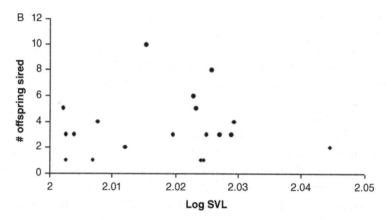

Fig. 2.5 *A*, A plot of sprint speed versus number of offspring sired, for collared lizards, *Crotaphytus collaris*. *B*, A plot of body size versus number of offspring sired, for collared lizards. Note the positive relationship between how fast lizards run and reproductive success. Thus, variation in sprint speed, not just variation in body size, explains a portion of reproductive success in these lizards. Taken from Husak et al (2006) with permission from Wiley-Blackwell Press.

words, it's good to be fast, but not too fast. The reasons for this penalty are unclear but could include correlated costs of extremely high performance, such as increased metabolic rate, deleterious behavioral traits, and so forth.

While it is easy to focus on studies that show significant selection, what about evidence for a lack of selection? It is noteworthy that, about half the time, researchers have found no significant selection on performance; that is, in many cases, variation in performance capacity among individuals is inconsequential in an evolutionary sense (Irschick et al. 2008). As explained in section 2.1, reasons for these results include the temporal variability of selection pressures, trade-offs with other traits that are more important to fitness, and the possibility that variation in some kinds of performance might be meaningless ecologically, at least when researchers were examining them.

2.6 Is selection on morphology stronger than for performance capacity?

A widely held view is that form should translate into function—in other words, morphological variation, such as the cross-sectional area of a muscle group, should translate into the increased ability to accomplish some functional task. If this view were true, then selection on morphology should be similar to selection on performance capacity. Another view is that selection should be stronger on performance than on morphology, as selection should act on performance primarily and on morphology secondarily. Consider again our hypothetical cheetah population, in which only the fastest individuals are able to capture prey and therefore survive and reproduce. As in most vertebrates, the length of the hindlimb predicts (with some error) maximum speed, and, therefore, fast cheetahs also tend to have long hindlimbs. A mark-recapture study shows that natural selection clearly favors fast cheetahs and, therefore, because speed and hindlimb length are inextricably linked, selection also favors long hindlimbs. So, if variation in hindlimb dimensions were the primary predictor of speed in cheetahs, then selection on hindlimb length and speed should be the same. But what if other factors were also important for dictating speed? Anyone watching the Olympics can see that there are many different body shapes competing in each event, and the "best-looking" athlete is not always the winner! In fact, factors such as motivation, technique (how body parts are used in space), and internal structural and physiological processes that may not be externally visible may also be important. In other words, one might expect that selection on performance should be direct and strong, whereas selection on morphology should be somewhat weaker.

Fortunately, there is a way to measure the intensity of selection, called the selection coefficient (beta), of which there are two primary kinds: one that measures the strength of directional selection (i.e., selection that favors only extreme [high or low] phenotypes or performance traits), and another that measures the intensity of stabilizing selection (selection that favors intermediate phenotypes or performance traits). Joel Kingsolver and his colleagues (2001) have compiled data on both kinds of selection coefficients from various selection studies of morphology (and other similar traits), and one can compare these values to recently compiled selection statistics for performance traits (Irschick et al. 2008). Contrary to the initial expectations set forth above, there is little difference in the strength of selection (either directional or stabilizing) between morphological and performance traits (fig. 2.6). In sum, there is no compelling

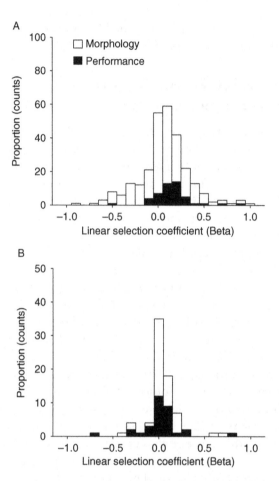

Fig. 2.6 *A*, A histogram of linear selection gradients from a variety of animal selection studies, showing the strength of directional selection on various performance traits. *B*, A histogram of quadratic selection gradients, showing the strength of stabilizing selection on performance traits. In both plots, black values are used for performance traits, whereas white values are used for other, morphological traits taken from Kingsolver et al. (2001). Data are from Irschick et al. (2008).

evidence that selection is either stronger or weaker on performance relative to morphological traits. However, this initial assessment should be viewed with caution. First, the total number of selection studies on morphology greatly exceeds that for performance studies, and more studies on performance traits are needed before a robust comparison can be made. Second, the most relevant comparison lies between selection on performance traits and the corresponding morphological traits that are directly linked to one another, such as in the above example of hindlimb length and maximum speed for cheetahs. Many of the morphological traits included in figure 2.6 may have little relevance for any performance trait and may even have little relevance for the animal in general. It would be valuable to compare the intensity of selection on performance traits to that on morphological traits that are known to influence performance. Another thing to keep in mind is that multiple categories of performance may be under selection simultaneously, and this may dilute the observed strength of selection on performance because of functional conflicts.

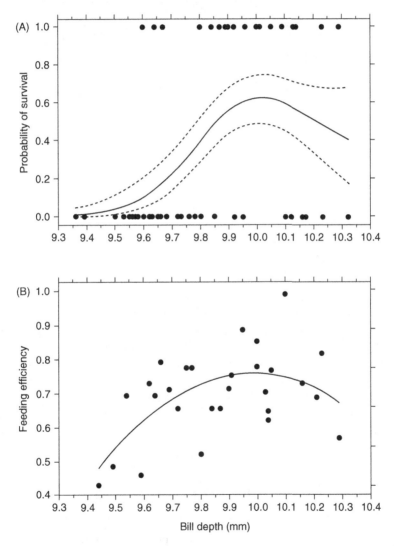

Fig. 2.7 *A*, A plot showing the fitness benefit for survival for crossbill birds with high values of bill depth. *B*, A plot showing the increase in feeding efficiency (up to a point) for crossbill birds with larger bill depths. *C*, A three-way plot showing different fitness peaks for crossbill birds for various combinations of bill depth and groove size of pine cones. Taken from Benkman (2003) with permission from Wiley-Blackwell Press.

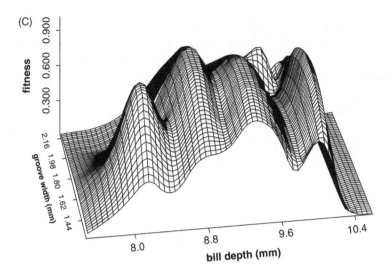

Fig. 2.7 (*continued*)

This might also result in comparable intensities of selection acting on performance and lower-level traits.

As an example of disconnect between selection on morphology and performance, there is significant selection on sprint speed in *Urosaurus* lizards (Irschick and Meyers 2007) but no significant selection on hindlimb length, despite a great deal of evidence that animals with long hindlimbs tend to run faster than animals with short limbs. Other such "disconnects" likely exist but have either never been published or never even been examined, because few studies examine selection on more than one variable in a given study. In fact, the "file-drawer" effect (so named because one tends to keep less exciting negative results tucked away in the file drawer) is a persistent problem that clouds our view of the nature of selection. Researchers will often publish exciting positive selection results but avoid publishing negative results for fear of criticism, or perhaps because of hope that more data in the future might change the findings.

2.7 Manipulative studies of animal performance and selection

While the majority of research on animal performance and fitness is conducted under "natural" circumstances in which animals are marked and left to fend for themselves in their natural habitat, such an approach is challenging in many cases. For example, many aquatic animals that move long distances are less constrained by physical barriers than terrestrial animals are and may be difficult to recapture. For such animals, researchers have used creative methods to artificially recreate the salient features of natural habitats, while preserving the ability to successfully measure fitness. For example, studies of smaller aquatic animals (some snakes, tadpoles, fish) often use cattle tanks that mimic small natural pools. Another aspect of manipulation involves changing not the physical environment in which the animals occur but the phenotype itself, with the ultimate aim of affecting performance capacity. The fins of swimming fishes have been manipulated in several studies, with variable outcomes. In a study of bluegill sunfish, ablating almost half of the pectoral fin did not result in a difference in braking performance but did change the way that braking performance was achieved (Higham et al. 2005). These fish abduct (move away from body) and protract (more forward) their fins during braking maneuvers, and the fish with reduced fins simply moved their fins faster, which ultimately exerted the same force on the water as larger fins moving slower.

Animal flight offers an opportunity to manipulate the phenotype and animal performance, because wing shape and area can be manipulated in some cases. Among flying animals, invertebrates are the simplest system in which to manipulate wing shape and size because many flying insects naturally suffer some damage to their wings. Butterflies are notable in possessing extremely large wings for their size. For example, butterflies possess low wing loadings (body mass divided by wing area) compared to other animals that use flapping flight. Unlike many flying animals, one can also remove portions of the wings of butterflies (which happens naturally because of damage they sustain) and then examine the effects of wing loss on fitness. Flight studies of butterflies in the laboratory indicate that a reduced wing loading (i.e., smaller wings per body mass) negatively influences both takeoff acceleration and lift production (Kingsolver 1999). In other words, butterflies with "clipped" wings are at a performance disadvantage compared to "control" butterflies with unclipped wings; but whether this manipulation affects their likelihood of survival is another question.

Surprisingly, wing manipulation did not affect survival in butterflies, as butterflies with clipped wings were just as likely to survive as butterflies with unclipped wings were. Why did this dramatic manipulation have no clear impact on fitness? First, even though reduced wings hamper flight performance, as long as butterflies are able to effectively move and evade predators, then there will be no fitness decrement. This underscores the point that evolution only proceeds to what is "good enough." So, why are butterfly wings much larger than needed for effective flight? The "overbuilding" of butterfly wings conforms to other structural and performance traits that are excessive or "overbuilt" relative to the everyday challenges that organisms face. Just as jet airliners have auxiliary engines in the case of the loss of one engine, animals frequently possess excessive structural capacity to protect against damage that might ensure from fights with other individuals, encounters from predators, or simply from wear and tear. Butterfly wings also serve another important function unrelated to flight, namely, thermoregulation. Most butterflies require a certain critical body temperature to fly and, by spreading their wings to capture the sun's rays, the butterflies use their wings as thermal heat collectors and therefore increase thermal heating. We don't yet fully understand the complex ways in which the shape of butterfly wings affects flight and thermoregulation, but suffice it to say that there are many influences on morphological shape besides the most obvious functional explanation (flight in this case).

As mentioned, another form of manipulation involves studying natural selection in enclosed environments in which animals can be easily marked and followed. Enclosure selection studies also offer the opportunity to examine the entire causal chain of events for how variation in morphology and performance capacity influences fitness. For example, while the interface between development and selection is largely neglected in field studies, researchers have been able to study them in enclosure experiments. The Oriental toad *Bombina orientalis* is a wonderful system in which to examine interrelationships among environmental factors (e.g., temperature), morphology, development, and performance capacity, because the toads easily breed in captivity and one can replicate the scale of their natural pond habitat using containers (Kaplan and Phillips 2006). If one exposes the eggs of this toad to varying temperatures, there is a direct effect on the phenotype, with higher lower temperatures resulting in longer tail lengths (despite slower growth rates) during the earliest developmental stages (tadpoles) as a result of delayed hatching. Longer tail lengths increase burst speeds and also the likelihood of survival during exposure to predators, which in this case were tadpoles from another frog species. Paradoxically, shorter tail lengths are also favored by natural selection, that is, selection on tail length is disruptive, and tadpoles with intermediate tail lengths are disfavored. Such alterations in maternal investment also play a role; too much maternal investment results in overly large eggs carrying offspring that are more likely to die because tadpoles from these eggs exhibit lower burst speeds. But, there is a catch. Higher temperatures can result in hatching at younger stages of development and with shorter tails, so the positive influences of higher temperatures on tail growth could be negated by the maternal burden of yolk that the mother provides. At higher temperatures at younger stages, even longer tails cannot push a big yolky gut very well, so increased maternal investment (i.e., egg size) results in decreased survival. This example is clearly a case in which one can observe the entire set of causal interactions among maternal investment, egg size, hatching times, developmental stages, and growth rates, as well as the effects of these interactions on fitness.

Manipulative selection studies have been used to evaluate the evolutionary consequences of predator-prey interactions. On the Caribbean island of Trinidad, David Reznick and his colleagues (1990) have conducted a long-term series of experiments examining how guppies, *Poecilia reticulata*, adapt to various natural predators. Guppies live in streams that differ in the type (species) and abundance of predators, with some guppy communities containing several kinds of ubiquitous dangerous predators (high predation), and others containing fewer and less predatory species. In transplant experiments, the researchers took guppies from high-predation sections of rivers and introduced them over barrier waterfalls that had excluded guppies and predators. In a different experiment, they transplanted predators over a waterfall that had excluded predators but not guppies. Did the guppies that had predators introduced into their neighborhood evolve faster escape speeds to survive in this dangerous new environment? By comparing natural high- and low-predation environments with the introduced populations, they found that the "experimental" fish possessed more rapid escape speeds for eluding predators (O'Steen et al. 2002). However, this finding did not exclude the possibility of phenotypic plasticity—the process by which organisms acquire or enhance their phenotypes, such as when our arms get larger because we do lots of pull-ups. Lab breeding experiments showed that this evolutionary change was not entirely plastic; it had some genetic basis and thus would not fully disappear if conditions were reversed. The ecological value of burst speed for Trinidadian guppies for eluding predators has also been shown in manipulative laboratory studies. When placed in artificial ponds with a large fish predator, the pike cichlid, *Crenicichla alta*, individual guppies with high burst speeds were more likely to survive encounters with the predator. Indeed, a difference of one standard deviation in burst speed resulted in a two- to threefold increase in survival probability for guppies in interactions with fish predators (Walker et al. 2005). In a further twist, the researchers also showed that the direction of the fish escape is crucially important for a successful escape. Their finding that performance traits can evolve rapidly when animals are exposed to novel environments was also bolstered by recent work showing that, over only a span of 36 years, Italian wall lizards, *Podarcis sicula*, in Croatia have evolved harder bites to consume harder prey (Herrel et al. 2008), although that study did not conclusively disprove the possibility of a plastic response.

Just as natural predators can act as agents of selection, one can mimic evolution by selective breeding, a process demonstrated with domestic cats and dogs. In a few cases, scientists have also used selective breeding to understand how performance capacity might evolve. One feature of locomotion that impacts locomotor endurance is the level of intermittency, or the number of "pauses" taken by animals during locomotion. Intermittent locomotion generates energetic savings compared to continuous running because it facilitates rapid recovery of ATP stores during intense aerobic or anaerobic exercise. By taking very small "breaks" (even a matter of seconds) during an otherwise exhausting bout of locomotion, animals and humans can increase the total amount of locomotor work accomplished prior to fatigue or exhaustion (a bit depressing if one plays basketball or soccer to build aerobic capacity!). Consequently, many animals that must move frequently "pause" often, presumably to avoid fatigue. Intriguingly, different individuals within a population vary in their degree of intermittency, thus setting the stage for selection to occur. When one breeds mice, *Mus domesticus*, for how much they run voluntarily on wheels, it takes only about 10 generations for a large difference to emerge (75% increase), with these high-runner mice ultimately (after about 23 generations) running about 2.7 times more revolutions/day, as compared to those from a group in which no selection occurred (Swallow et al. 1998; Careau et al. 2013). In addition, the high-runner mice evolved to run more intermittently, with many pauses (Girard et al. 2001). These studies also revealed a strong genetic basis to wheel-running activity (about 28% of the variation among individuals in the original starting population). However, that's not all! Other studies revealed that lines of mice bred for high levels of voluntary wheel running also exhibited alterations in their brains compared to control mice, and these former mice displayed features much like those found in humans with attention deficit hyperactivity disorder. Drugs such as Ritalin appear to reduce the high-intensity running for mice bred for voluntary running, whereas control mice showed little sensitivity to the drug (Rhodes et al. 2005)!

This less natural, though fascinating, approach to selection has also been applied to fruit flies (*Drosophila*), with remarkable results (Marden et al. 1997). Using computerized tracking of three-dimensional movements of flies, a research team showed how selection altered the flight abilities of fruit flies. After 160 generations of selection for the ability to fly "upwind" (upward), the selected group had significant increases (relative to the control group) in mean flight velocity, as well as alterations in their angular trajectory. Overall, these data indicate that selection for flight performance can alter the frequency of phenotypes capable of attaining high levels of flight performance, but there was little change in change of population level maximal performance. Other studies, also with *Drosophila melanogaster*, showed that increased selection on flying speeds in a wind tunnel, resulted in a substantial increase in mean flight speed (relative to the control), from 2 cm/s to 170 cm/s. These laboratory selection studies reveal that, at least in some animals, there is room to improve locomotor performance through human-induced selection. To what extent this pattern would hold up in wild populations of animals would be an interesting topic for future research.

2.8 Complex forms of selection on performance

Up to this point, most of the discussion has centered on the idea that selection favors only a single kind of performance. However, are there complex forms of selection that favor multiple kinds of performance traits, perhaps driven by the presence of alternative phenotypes within animals? Documented examples of complex forms of selection often arise from studies of the feeding structures of animals, although these studies have focused on morphology, with some indirect links to performance. In African finches, *Pyrenestes ostrinus*, there are two alternative beak types, each of which is adapted to specialize on different kinds of prey. Those birds with larger beaks can more quickly consume harder seeds, compared to birds with smaller beaks, and this difference seems to have a genetic basis (Smith 1987, 1993). In crossbill finches, *Loxia curvirostra*, in North America, a similar evolutionary dynamic exists. Crossbill finches use their scissor-like beaks to pry open pine cones to consume the seeds hidden within the cones (Benkmen 2003). Because there are several kinds of pine cones, and because each kind differs in how easy or hard it is to pry open, beak phenotypes differ in their effectiveness for feeding on different types of pine cones (fig. 2.7). Consequently, within crossbills, there is a complex fitness structure that is related to the physical shape of the pine cones (fig. 2.7).

Not all examples of "complex" selection are restricted to feeding. During escape from predators, rapid locomotion is an important escape strategy for many animals, but the decision for how and when to use rapid locomotion is dependent on various factors such as the type of predator, and the intrinsic performance abilities of the individual in question (i.e., is the individual fast or slow?). In some populations of garter snakes, genus *Thamnophis*, there is another factor: dimorphism in the back color pattern, with some individuals possessing a striped color pattern, and others having a more checkered or mottled appearance (Brodie 1992). The presence of stripes seems to pose challenges for visual predators because the stripes create the optical illusion that the snake is traveling faster than it really is. Moreover, garter snakes sometimes perform "reversals" in which they suddenly reverse their direction of travel during escape, presumably to allow them to blend into the habitat, which is a strategy favored by mottled or checkered snakes. Mark-recapture studies by Ed Brodie Jr. show that the likelihood of survival in juvenile garter snakes is a consequence of whether the snake is mottled or striped, and the likelihood of reversals. Mottled snakes are at a fitness advantage if they employ reversals, whereas striped garter snakes are not (fig. 2.8). This phenomenon of traits being favored in discrete combinations is called correlational selection and, in many cases, these favored combinations of traits also are genetically correlated, that is, they can evolve in tandem; however, whether genetic correlation ultimately leads to correlational selection is less clear (Delph et al. 2011; Roff and Fairbairn 2012).

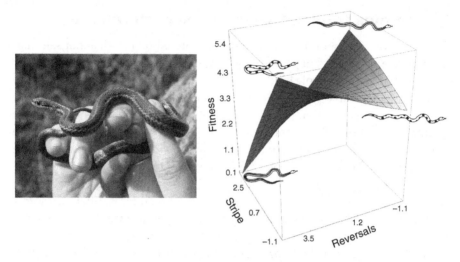

Fig. 2.8 *Left*, Image of a garter snake (*Thamnophis* sp.), taken from Wikimedia Commons. *Right*, A three-way plot showing relationships between stripe pattern, reversals during escape, and fitness in garter snakes. Taken from Brodie (1992) with permission from Wiley-Blackwell press.

References

Arnold SJ. 1983. Morphology, performance, and fitness. American Zoologist 23:347–361.

Artacho P, Nespolo R. 2009. Natural selection on reducing energy metabolism in the garden snail (*Helix aspersa*). Evolution 63:1044–1050.

Benkman, CW. 2003. Divergent selection drives the adaptive radiation of crossbills. Evolution 57:1176–1181.

Brodie ED III. 1992. Correlational selection for color pattern and antipredator behavior in the garter snake *Thamnophis ordinoides*. Evolution 46:1284–1298.

Bumpus HC. 1899. The Elimination of the Unfit as Illustrated by the Introduced Sparrow, *Passer domesticus*. Biological Lectures, Marine Biology Laboratory, Woods Hole, MA.

Careau V, Wolak ME, Carter PA, Garland T Jr. 2013. Limits to behavioral evolution: The quantitative genetics of a complex trait under directional selection. Evolution 67:3102–3119.

Delph LF, Steven JC, Anderson IA, Herlihy CR, Brodie ED. 2011. Elimination of a genetic correlation between the sexes via artificial correlational selection. Evolution 65:2872–2880.

Endler JA. 1986. Natural Selection in the Wild. Princeton University Press, Princeton, NJ.

Garland T Jr., Losos JB. 1994. Ecological morphology of locomotor performance in squamate reptiles. In Wainwright P, Reilly SM, eds. Ecological

Morphology: Integrative Organismal Biology. University of Chicago Press, Chicago, IL, pp. 240–302.

Girard I, McAleer MW, Rhodes JS, Garland T Jr. 2001. Selection for high voluntary wheel-running increases speed and intermittency in house mice (*Mus domesticus*). Journal of Experimental Biology 204:4311–4320.

Herrel A, Huyghe K, Vanhooydonck B, Backeljau T, Breugelmans K, Grbac I, Van Damme R, Irschick DJ. 2008. Rapid large scale evolutionary divergence in morphology and performance associated with the exploitation of a novel dietary resource in the lizard *Podarcis sicula*. Proceedings of the National Academy of Sciences 105:4792–4795.

Higham TE, Malas B, Jayne BC, Lauder GV. 2005. Constraints on starting and stopping: Behavior compensates for reduced pectoral fin area during braking of the bluegill sunfish *Lepomis macrochirus*. Journal of Experimental Biology 208:4735–4746.

Husak JF. 2006. Does speed help you survive? A test with collared lizards of different ages. Functional Ecology 20:174–179.

Husak JF, Fox SF, Lovern MB, Van Den Russche. 2006. Faster lizards sire more offspring: Sexual selection on whole-animal performance. Evolution 60:2122–2130.

Irschick DJ, Meyers JJ. 2007. An analysis of the relative roles of plasticity and natural selection on morphology and performance in a lizard (*Urosaurus ornatus*). Oecologia 153:489–499.

Irschick DJ, Meyers JJ, Husak JF, Le Galliard JF. 2008. How does selection operate on whole-organism functional performance capacities? A review and synthesis. Evolutionary Ecology Research 10:177–196.

Kaplan RH, Phillips PC. 2006. Ecological and developmental context of natural selection: Maternal effects and thermally induced plasticity in the frog *Bombina orientalis*. Evolution 60:142–156.

Kingsolver JG. 1999. Experimental analyses of wing size, flight and survival in the western white butterfly. Evolution 53:1479–1490.

Kingsolver JG, Hoekstra HE, Hoekstra JM, Berrigan D, Vignieri SN, Hill CE, Hoang A, Gibert P, Beerli P. 2001. The strength of phenotypic selection in natural populations. American Naturalist 157:245–261.

Lande R, Arnold SJ. 1983. The measurement of selection on correlated characters. Evolution 37:1210–1226.

Lappin AK, Husak JF. 2005. Weapon performance, not size, determines mating success and potential reproductive output in the collared lizard (*Crotaphytus collaris*). American Naturalist 166:426–436.

Marden JH, Wolf MR, Weber KE.1997. Aerial performance of *Drosophila melanogaster* from populations selected for upwind flight ability. Journal of Experimental Biology 200: 2747–2755.

Miles DB, Sinervo B, Frankino WA. 2000. Reproduce burden, locomotor performance, and the cost of reproduction in free-ranging lizards. Evolution 54:1386–1395.

O'Steen S, Cullum AJ, Bennett AF. 2002. Rapid evolution of escape ability in Trinidadian guppies (*Poecilia reticulata*). Evolution 56:776–784.

Reznick DN, Bryga H, Endler JA. 1990. Experimentally induced life-history evolution in a natural population. Nature 346:357–359.

Rhodes JS, Gammie SC, Garland T Jr. 2005. Neurobiology of mice selected for high voluntary wheel-running activity. Integrative and Comparative Biology 45:438–455.

Roff DA, Fairbairn DA. 2012. A test of the hypothesis that correlational selection generates genetic correlations. Evolution 66:2953–2960.

Schluter D. 1988. Estimating the form of natural selection on a quantitative trait. Evolution 42:849–861.

Smith TB. 1987. Bill size polymorphism and interspecific niche utilization in an African finch. Nature 329:717–719.

Smith TB. 1993. Disruptive selection and the genetic basis of bill size polymorphism in the African finch *Pyrenestes*. Nature 363:618–620.

Swallow JG, Carter PA, Garland T. Jr. 1998. Artificial selection for increased wheel-running behavior in house mice. Behavior Genetics 28:227–237.

Walker JA, Ghalambor CK, Griset OL, Kenney DM, Reznick DN. 2005. Do faster starts increase the probability of evading predators? Functional Ecology 19:808–815.

Wilson AM, Lowe JC, Roskilly K, Hudson PE, Golabek KA, McNutt JW. 2013. Locomotion dynamics of hunting in wild cheetahs. Nature 498:185–189.

3 The ecology of performance II
Performance in nature

3.1 Why study animal performance in nature?

Most animal performance studies are conducted under "standardized" laboratory conditions, and for good reason. To understand the factors (e.g., food
intake, temperature) that affect performance, some level of control is necessary, and laboratory studies offer superior control over such environmental
variables, compared to field studies. For instance, if one is interested in the
effects of temperature on vocalization, one can precisely control temperature
by using temperature chambers, while controlling for other potentially confounding environmental variables such as diet, age, and so forth. To test basic
hypotheses regarding the role of energetic and neuromuscular mechanisms on
performance, bulky and sophisticated (and expensive!) equipment is often necessary, and such equipment is not suitable for use in nature, although technological advancements are rapidly changing these limitations. Scientists now
have available to them mobile devices that provide remarkable access to the
hidden world of animals. For example, mobile monitors can measure, among
other features, heart rate, water depth, body temperature, body orientation, location on the earth via GPS, and even certain muscle movements, from animals
as diverse as cheetahs, sharks, turtles, and marine mammals (fig. 3.1)! Further,
accelerometers, which measure both velocity and acceleration instantaneously,
now enable scientists to measure how fast animals move during a variety of
tasks. These devices are getting smaller and smaller, with the ability to measure the dynamics of motion of very small animals (<10 g) fast approaching.

Even beyond technological advances, there are important conceptual reasons for studying animal performance directly in nature (Turchin 1998). First,
with the exception of some domestic animals, most animals perform in the

Animal Athletes. Duncan J. Irschick and Timothy E. Higham. © Duncan J. Irschick and Timothy E.
Higham 2016. Published in 2016 by Oxford University Press.

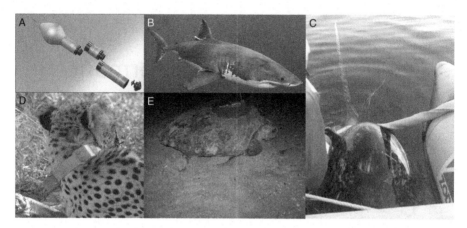

Fig. 3.1 *A*, A typical field-portable satellite tag that allows transmission of positional data, and potentially other data when combined with other sensors such as accelerometers, temperature loggers, and so forth. *B–E*, Examples of animals that are now studied in the wild using remote-sensing equipment. *B*, A shark. *C*, A narwhal. *D*, A cheetah. *E*, A sea turtle. All images from Wikimedia Commons except for the cheetah image, which is taken from Wilson et al. (2013) with permission from Nature Publishing group.

laboratory under duress and, while this factor may be unimportant for certain physiological tests, it might influence our ability to determine the true ecological factors that limit performance. Consider the majestic elephant seal, which regularly dives to very deep depths (up to 500 m) to hunt for food and has evolved remarkable abilities to hold its breath and hunt under extreme physical duress. Humans currently lack the capability to replicate these astounding diving events under artificial conditions, and it is well known, from observing voluntary diving events, that responses to forced submergence vary dramatically. Moreover, even if we could design a holding tank 500 or more meters deep, such artificial tests might not reveal much about how these animals dive. This is because "voluntary" dives in nature occur on the animals' own time schedules, and under their own constraints. Studies with dolphins show that, when voluntarily diving even for relatively short distances, there is a striking difference in their diving physiology in manmade versus natural environments. In other words, not only may an animal perform differently under forced laboratory conditions than in its natural habitat, but the way that its body responds to these physical challenges may also differ. Just as in humans, motivation in animals needs to be considered when determining how well they do a certain task. Second, while laboratory studies might reveal *how* animals can accomplish certain performance tasks, such as how muscles and tendons provide the driving force for locomotion, they reveal little about *why* these animals perform these tasks. Thus, by studying animals in nature, we can gain insight into the evolutionary and ecological pressures that have molded their performance capacities.

Another difference between laboratory studies and those conducted in nature is that the former often examine a single behavior (e.g., maximum sprinting speed). In reality, animals perform a multitude of complex behaviors throughout the day. In addition, combinations of behaviors may be necessary to accomplish something. For example, a kangaroo rat may jump (up to 9 feet!) away from a snake's strike but then dash to a bush upon landing. Thus, both jumping and hopping quickly are likely important movements during an escape. The combination of these may be very relevant, but laboratory studies typically focus on a single movement.

3.2 What is ecological performance?

Ecological performance is performance as measured in an animal's natural surroundings (Irschick and Garland 2001). To provide a better understanding of what is meant by ecological performance, consider an example of an insect that has a maximum flight speed of 0.5 m/s, which can be determined using the most current methods for assessing flight, such as with a flight tunnel. However, by placing a small accelerometer on the insect when it is flying around naturally, biologists can study its actual (ecological) performance. For the sake of this example, let's say that we find that this insect, on average, flies at around 90% of its maximum capacity when eluding a key predator, around 70% when chasing a common prey item, and only 10% when moving between perching sites. While not exhaustive, these numbers inform us about how fast this insect flies under a variety of different circumstances. Cumulatively, we can consider all of these values as being roughly representative of the ecological performance capacities of this theoretical insect. One could extrapolate this approach to nearly every organism or performance metric, although the dichotomy between field and laboratory measures will be more important for some species and traits than others. For example, one could examine biting and compare the maximum force an animal can exert with its jaws to the actual force used when the animal is eating. Alternatively, one could compare the maximum swimming speeds of fish to the actual speeds these animals use in nature. Many other possible examples exist.

Just as with measurements of laboratory performance, there are several caveats. First, to interpret ecological performance, one should ideally first measure maximum performance under standardized conditions, to provide a "top" baseline against which to compare ecological performance. In some rare cases, it will prove impossible to measure maximum performance in the laboratory, and animals may only make full use of their performance capacities in nature. Second, in some cases, there may be little difference between standard laboratory and field measurements of performance capacity, especially if the performance trait is dictated by simple biomechanical relationships with morphological traits. On the other hand, if the performance trait under study is affected by behavior, then there may be a marked difference between "laboratory" performance and "ecological" performance. For instance, the degree of locomotor performance is greatly affected by behavior, which can be determined by, for example, an animal's relative level of motivation. This is clearly evident in frog jumping competitions, in which humans motivate their competing frogs to jump as far as they can, by yelling and moving quickly toward them. These jumping contests occur throughout the United States, with a very famous event held annually in Calaveras Country (since 1928). Of course, the owners of the frogs are also very motivated—by the cash prize for longest jump! Let's examine some examples of how animals perform in nature and explore these issues further.

3.3 Impact of behavior on ecological performance

Performance capacities have rarely evolved for use during only one behavior and, because of their differing demands, performance capacities will often differ among behaviors. To some extent, performance differences among behaviors may be voluntary and may represent different levels of "effort" for different tasks. For instance, fleeing from a predator will typically require a lot more effort than simply moving around one's habitat when undisturbed. Studies of flight in birds and insects have provided some of the best data examining the impact of behavior on different kinds of locomotor performance in nature. Flight can be energetically expensive; yet birds are capable of flying extraordinarily long distances, often without stopping to rest. For example, arctic terns fly distances over 3,000 km long and can fly over the entire Atlantic Ocean without stopping—an impressive feat for a human-made plane, let alone for a small animal! By using various means of quantifying how fast birds fly in nature and then using theoretical models for determining the relationship between speed and energy spent, researchers have been able to determine whether birds can "optimize" their flight speeds to maximize energetic output, in much the same way that planes fly at speeds that are optimal for increasing gas mileage (Hedenström and Alerstam 1992; Pennycuick 1997; Pennycuick et al. 2013). Because of the complex manner in which speed and energy are related, researchers have tested several hypotheses. One idea is that birds maximize the distance traveled per unit energy expended, also called the maximum range speed (V_{mr}) hypothesis. If true, then birds should fly at speeds in nature that would result in high ratios of speed to power. Remember that force equals mass times acceleration, and power equals force times speed. Therefore, total power production in an animal can be viewed as the ability of animals to produce force using their muscles.

Another idea is that birds should fly at speeds that minimize the amount of power they exert as they fly, known as the minimum power speed (V_{mp}) hypothesis. Unlike the former hypothesis, this one predicts that birds should fly at speeds that will minimize the ratio of power to speed. One caveat to consider is that birds, like any animal, may not always be interested in flying optimally, especially if energy is not limiting. Scientists continue to test these models, but one conclusion so far is that how large a bird is influences how much power it can recruit from its muscles. In general, smaller birds usually fly at speeds that are higher than or equal to the predicted V_{mp} or V_{mr}. On the other hand, larger birds usually have flying speeds that are lower than predicted speeds (Hedenström and Alerstam 1992; Welham 1994; Pennycuick 1997; see fig. 3.2). The reasons for this trend are unclear, but it's possible that small birds may be able to obtain greater amounts of power from muscles used for flight than larger birds can (Pennycuick 1997). Also, larger birds may seem to fly more slowly because they more frequently "slope-soar" (i.e., fly up and down in a rhythmic fashion, thereby saving energy because of the use of gliding) than small birds do. This flying method creates certain constraints because, as anyone going downhill can testify, it's hard to control one's speed under such conditions.

For birds, different styles of flying will influence how fast they fly in nature. Birds are capable of moving their wings in different ways during flight, resulting in flapping flight, hovering, gliding, and even diving. Birds will also fly in a back-and-forth motion when soaring above the waves. The different kinds of flight seem to differ in speed. Field studies show that birds achieve higher field speeds during flapping flight compared to other forms of flight. Not surprisingly, diving, which is used by gyrfalcons, is the fastest of all kinds of natural flight, and birds that use this kind of flight enhance their performance by modifying their wing and body posture to minimize drag. Birds will also sometimes employ erratic flight behavior, which can markedly reduce speed, resulting in a performance difference of as much as 29% (Blake et al. 1990).

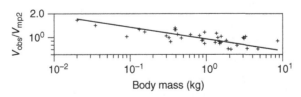

Fig. 3.2 A plot of body mass versus the ratio of observed velocity to minimum power velocity, for a variety of bird species. Note the negative relationship between the variables; this relationship suggests that smaller birds may be able to obtain more power from their muscles for flying, compared to larger birds; V_{obs}, observed velocity; V_{mp2}, minimum power velocity. Redrawn from Pennycuick (1997) with permission from the Company of Biologists Ltd.

Predation is one of the most powerful influences on how fast animals will move in their natural environment. The behavior of eluding predators usually dictates high levels of locomotor performance, unless a species relies on other aspects, such as unpalatability, to avoid being eaten. Butterflies differ greatly in their level of palatability to prevent being eaten, with some species using toxic secretions garnered from their consumption of plants. Other species are more palatable and have to rely on other means to elude predators. Robert Dudley and his colleagues have examined the flight of different species of butterflies, ranging from unpalatable to palatable species. In general, palatable butterflies fly at faster natural speeds and are more skillful at eluding predators in cage experiments. The morphology of palatable predators appears to enhance their flight performance, as their center of mass is positioned near their wing base. Unpalatable butterflies are generally slower fliers and cannot easily elude predators (Dudley and Srygley 1994).

Research with accelerometers and satellite tags from large marine animals such as sea turtles and whale sharks show that these animals vary how fast they move in nature, depending both on their reproductive state and on other behavioral factors. Research on loggerhead sea turtles, *Caretta caretta*, showed that these turtles moved more often during the reproductive season than during the nonreproductive season, possibly because females interacted more with males during the reproductive season than during the nonreproductive season. Turtles also were more active in colder water, likely because they were trying to find patches of warm water (Fossettee et al. 2012). Similar work with whale sharks, *Rhincodon typus*, showed that these large gentle beasts varied their energy use during dives, depending on the angle of their dives (Gleiss et al. 2011; fig. 3.3). Descending dives required less estimated energy, as compared to ascending dives, and when the whale sharks ascended or descended at steeper angles, the energetic requirements of movement went up. This result shows that the geometry of movement influenced energetics of movement in nature and also that variation in energetic use likely reflected different ecological requirements.

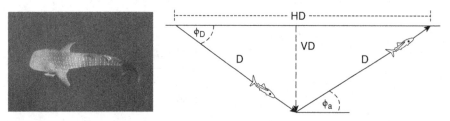

Fig. 3.3 *Left*, A whale shark, *Rhincodon typus*. *Right*, Diagram of the variables measured for whale sharks during free locomotion in nature by Gleiss et al. (2011). The variables measured were ascent pitch, (φ_a), decent pitch (φ_D), distance (D), horizontal distance (HD), and vertical distance travelled (VD) within a dive. Taken from Gleiss et al. (2011) with permission from Wiley-Blackwell press.

3.4 Environmental impacts on ecological performance

In the laboratory, human scientists can isolate and study the effects of several environmental variables to determine their impact on performance abilities. However, in nature, humans usually cannot control the environment in the same way. Further, animals typically must cope with a wide array of variable conditions, including temperature, wind, humidity, and so forth. Therefore, it would be useful to understand how animals can perform proficiently in the presence of such variability. One example is the impact of wind on the flight of birds. Anyone who has flown in a plane and encountered turbulence has realized that wind can have a negative impact on how fast an object can fly—or how your stomach feels! Moreover, favorable tail winds can greatly enhance flight speed in flying objects. There is a substantial amount of data showing that flight speeds of animals are higher when they can use tail winds to their advantage (e.g., McLaughlin and Montgomerie 1985; Wakeling and Hodgson 1992) and, therefore, just as airplane pilots must plan for, and account for, wind speeds to save energy and time, birds must do the same. As an example, in turkey vultures, much of the variation in their flight speed, or underlying physiological measurements (e.g., heart rate), is dependent on the wind patterns, and these birds are fully capable of increasing their gliding speeds with little additional investment of energy (Mandel et al. 2008). If only we could move around so easily and so freely!

Field studies allow us to understand the rich interaction between the environment and performance in a manner that cannot be duplicated in the laboratory. Anyone who has tried running up a hill knows that it requires a lot more energy than running on a flat surface does! This also holds true for nonhuman animals, but how much more energy is expended on inclines varies according to the size of the animal. Let's break this down. Potential energy is related to the height (of an object or animal) relative to a baseline height. In order to move up an incline, an animal must increase the stored potential energy. This increase in potential energy is equivalent to the work needed to move that object up to that point. Therefore, because of the increase in potential energy moving up inclines, more work must be done. Although all animals will face an increase in work to move uphill, the challenge diminishes as animals become smaller. While this fact is well known, it is less understood how animals perform on inclines in nature and whether they might even use them as part of their escape strategy, such as when a small animal uses running up a hill to elude a large animal. In the Southwestern region of the United States, the zebra-tailed lizard, *Callisaurus draconoides*, is normally a denizen of desert washes, but in some areas it occurs on the edge of sand dunes, where it moves on compliant sand. While challenging for humans to move on, these surfaces offer a unique opportunity to study ecological performance, because these lizards will leave their footprints in the sand. Because of the straightforward relationship between stride length and speed in most limbed vertebrates, one can estimate the speed of movement from these tracks. A straightforward prediction is that, because moving up inclines is energetically expensive, lizards should avoid moving directly on them. However, upon closer inspection, this prediction is overly simplistic (Irschick and Jayne 1999). Although lizards show no tendency to run up shallow (<15°) inclines (relative to flat surfaces), they do show a propensity to run up very steep (>15°) hills. Specifically, they prefer to run up about half the hill and then run alongside it, often for long distances. The value of such an escape strategy seems clear. By running alongside a very steep hill, the small lizard can effectively keep large predators from capturing them. It is also worthwhile to note that these lizards will run at nearly 90% of their maximum speed in nature (fig. 3.4); this observation explains why, since they can run up to almost 6 m/s, we could never outrun them!

What goes up must come down; but few people have studied "downhill" locomotion. Some animals might have tricks for getting back down, such as jumping, gliding, or simply falling. However, most animals must actually climb down if they climbed up in the first place (domestic cats in trees might be an exception here!). Rather than do work to push its body up, now the animal must stop itself from tumbling down. Thus, issues related to stability and deceleration are critical for avoiding a potentially catastrophic injury. Thus, studies that quantify how animals move both up *and* down in their natural habitat are necessary for fully understanding how animals move on inclines.

For animals that live in the open ocean, the challenges of diving are significant, especially for animals that dive to deep depths (>50 m). As water depth increases, water pressure increases and temperature decreases. Further, for air-breathing animals, such as marine mammals, a deep dive means a longer time without access to fresh oxygen. Marine mammals are perhaps the best known of all diving animals, and their physiology and diving performance in nature have been well studied. Most marine mammals seem to exhibit a classic physiological response to diving. This response includes a diminished heart rate (sometimes as low as 4 beats/min; fig. 3.5), which is correlated with a diminished metabolic rate and thus a lesser need for oxygen; constriction of peripheral blood vessels, thus reducing heat loss; and an overall shunting of blood flow from the periphery to the internal organs (Andrews et al. 1997; Davis et al. 2004). The typical behavior for diving involves a rapid descent, slow horizontal movements at the bottom of the dive, and then a rapid ascent. In general, for deeper dives, marine mammals move at faster rates; however, the relationship between water depth and ascent and descent speed is not linear but curvilinear.

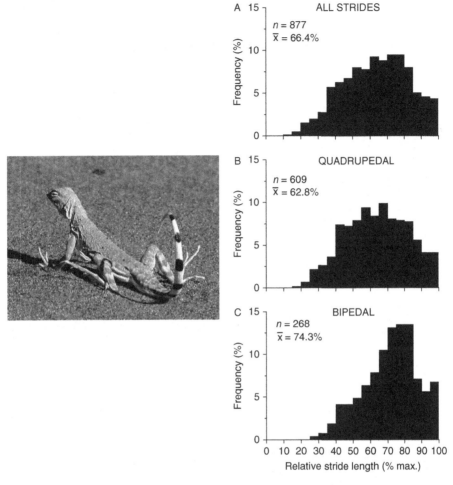

Fig. 3.4 *Left*, A zebra-tailed lizard, *Callisaurus draconoides*. *Right* (*a–c*), Histograms of stride lengths made by the when escaping from a potential threat on sand dunes in southern California, USA. Values are scaled as a percentage of maximum and indicate that lizards frequently move at very high percentages of maximum speed and that stride lengths during bipedal locomotion are about 12% greater than during quadrupedal locomotion. Taken from Irschick and Jayne (1999) with permission from the University of Chicago press. Image of zebra-tailed lizard taken from Wikimedia Commons.

There are a myriad of challenges that fishes face in their natural habitat, including crashing waves in intertidal regions, natural fluctuations in light levels, rapid temperature fluctuations (especially in isolated tide pools), very fast ambient flow conditions in rivers, and turbid conditions as sediment enters flowing water (Higham et al. 2015). Anyone who has walked along a shore with crashing waves can attest to the fact that such variables can create a harsh environment for making a living. Nonetheless, few studies have examined the

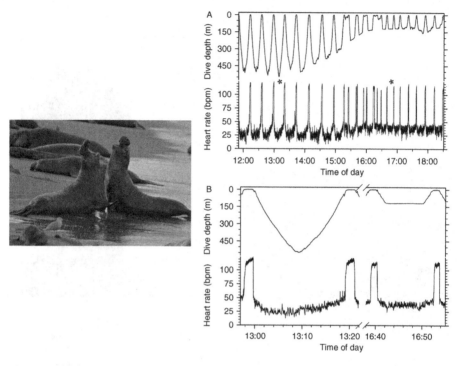

Fig. 3.5 *Left*, Northern elephant seals, *Mirounga angustirostris*. *Right*, Plots of dive depth and heart rate during a dive of a northern elephant seal equipped with remote-sensing equipment. The bottom plot shows a close-up of the two time periods marked with asterisks. Note the very deep depths and extremely low heart rate for the seal as it moves along the bottom. Image of elephant seals taken from Wikimedia Commons. Graphs taken from Andrews et al. (1997) with permission from the Company of Biologists Ltd.

importance of these variables on the motion and prey-capture ability of fishes. Luckily, fishes have a number of control surfaces (body and fins) that can help mitigate the negative impacts of destabilizing forces (Webb 2006). It is clear that much work is needed to fully understand how environmental variables impact aquatic animals, especially given the rapid changes that are occurring because of climate change and physical alterations caused by human activity.

Another environmental aspect that is challenging, if not impossible, to reconstruct under laboratory conditions is predator-prey interaction among relatively large animals. As we often see on the Discovery Channel, hunting prey and escaping predators can occur over large distances, whether on land, in air, or under water. However, whereas small animals (e.g., fish, lizards) can be manipulated in laboratory studies, larger animals, such as sharks or cheetahs, cannot be easily studied in the lab. Nonetheless, Alan Wilson's group at the Royal Veterinary College in London found an ingenious solution for studying how cheetahs hunt. They created a wireless collar that combined GPS tracking technology with an inertial measurement unit (IMU) that had a gyroscope, a magnetometer, and an accelerometer, thus enabling them to calculate measurements of both position and velocity when the cheetahs moved about (Wilson et al. 2013). Cheetahs are typically portrayed as sprinting at top speed after some frightened animal, and the data do reveal that cheetahs can reach impressive top speeds—a peak of 25.9 m/s; however, the data also show that cheetahs rarely obtain such high speeds and instead often hunt and capture prey at submaximal speeds (fig. 3.6). In fact, these data show that cheetahs may be built more for acceleration and maneuvering than straight-line speed and that hunting strategy may be an important component of effectively capturing prey.

Predators and prey exist in a dynamic struggle in which each is likely to modify their behavior in relation to the other. Emerging tracking data are increasingly revealing this remarkable modulation of behavior in nature. A recent example comes from satellite-tracking studies of bull sharks, *Carcharhinus leucas*, and tarpon, *Megalops atlanticus*, in southern Florida (Hammerschlag et al. 2012). Bull sharks are predators of tarpon, and these two species both occupy the warm coastal waters around Florida. The satellite-tracking data reveal a "landscape of predation" in which a high incidence of bull sharks can

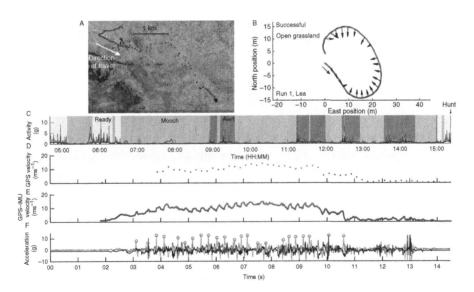

Fig. 3.6 Positional data during an 11 h period, taken from a cheetah equipped with a custom-made remote-sensing collar (Wilson et al. 2013). *A*, Overhead view of the habitat. *B*, The cheetah's movement during the period of observation. *C*, Histogram of the cheetah's activity pattern during the period of observation. *D–F*, Performance parameters for a cheetah chasing a prey item. *D*, Velocity calculated using GPS data only. *E*, Velocity calculated using a combination of GPS data and data recorded by the inertial measurement unit (IMU) in the cheetah's collar. *F*, Acceleration. Note that velocities are well over 10 m/s for portions of the chase, and acceleration values exceed 10 g.

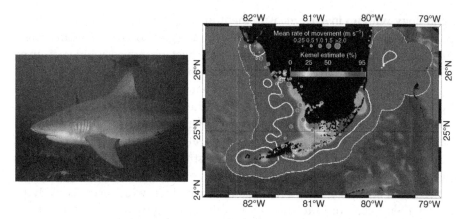

Fig. 3.7 *Left*, Bull shark, *Carcharhinus leucas*. *Right*, Map showing the relative likelihood of bull shark abundance as well as the abundance and movement of tarpon, *Megalops atlanticus*, around the coastal areas of southern Florida. Both bull sharks and tarpon, which bull sharks consume, were satellite tagged, and tarpon movements were examined in the context of shark distribution. The contours represent accumulative kernel density values for bull sharks (thus representing the likelihood of encountering a bull shark), whereas the circle sizes indicate tarpon locations as well as rates of movements. Tarpon were found to move faster through areas with higher likelihoods of encountering a bull shark. Bull shark image from Wikimedia Commons. Graph taken from Hammerschlag et al. (2012).

be viewed as being particularly dangerous for tarpon. In fact, the satellite data show that tarpon will move quickly through areas of high bull-shark abundance, and more slowly through less "dangerous" areas (fig. 3.7), thus revealing that tarpon likely "perceive" dangerous areas and adjust their ecological performance accordingly.

Finally, while it would not normally be thought of as an "environmental" variable, the population of origin is a potentially important factor that can influence decisions on how far and how fast animals will move around their habitat. Because populations differ in the number of individuals they contain, there are often differences among populations in overall morphological and genetic diversity, and such differences can influence behavior. This phenomenon was observed in a study in which Ovaskainena et al. (2008) used radar to track individually marked butterflies. The researchers showed that butterflies originating from newly established small populations moved more often than those originating from older (>5 years old) small populations or large continuous populations (fig. 3.8). The reasons underlying this result are complex but may relate to the need of butterflies in small, newly established populations to move longer distances than butterflies in larger or more well-established populations would have to in order to find appropriate food, resting sites, and reproductive opportunities.

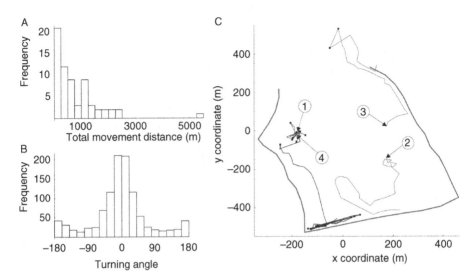

Fig. 3.8 Movement of individual Glanville fritillary butterflies, *Melitaea cinxia*. Individual butterflies originated from populations that were small and newly established, small and well established (>5 years old), or large and continuous. *A*, Movement distances. *B*, Turning angles. *C*, Examples of differing kinds of butterfly paths. The starting locations are represented by numbered arrows for four different butterflies (1–4). The black dots represent moments when the butterflies turned back (i.e., when the turning angle was greater than 135°). Taken from Ovaskainena et al. 2008 with permission from the National Academy of Sciences.

It is clear that much of our knowledge about performance capacity is based upon animals that are amenable to such studies in the laboratory. This snapshot of animal performance may be misleading; nonetheless, the above studies show how scientists can quantify a range of variables on a range of animals with great rigor, thereby opening the door for a more holistic field that integrates typical "laboratory" measures on animals in nature.

3.5 What percentage of maximum capacity do animals use in nature?

When we watch the human Olympics, we are pretty confident that everyone is at least trying fairly hard. The combination of overwhelming media attention with the promise of riches and fame makes a great motivator! While there have almost certainly been suboptimal efforts in the history of the Olympics, they are likely rare compared to the all-out efforts. For animals, the degree of effort an animal puts into a task is more complicated. The most pressing issue is whether exerting effort will increase fitness for the animal. Will running at maximum speed allow an animal to flee a predator or capture an elusive prey item? Or will doing so place the animal at risk because of other factors, such as the fact that using maximum effort may waste valuable energy or compromise an individual's ability to defend a territory? There is an emerging body of work that allows us to address these questions.

Several studies show that, when eluding a potential predator, lizards will regularly employ around up to 90% of their maximum capacities. By comparison, when engaging in other activities such as feeding or moving about undisturbed, this value drops considerably to between 30% and 70% (Irschick and Losos 1998). The percent of maximum capacity utilized may also depend on the structure of the habitat. For example, Namibian geckos living in rocky habitats use a lower percentage of their maximum sprinting capacity when escaping predators than related species that live in open, flat habitats do (T. Higham, unpublished data). This result may be related to the lizards having a reduced need to escape with high speeds when living in a complex habitat that provides a multitude of refuges.

There are also examples of how animals rarely reach their maximum speeds in nature. Gyrfalcons, *Falco rusticolus*, are potentially able to fly up to 170 m/s; however, in trials with live animals, such high speeds were rarely reached, either because it was overly challenging for the birds to do so or because it posed some severe risks to them to do so (Tucket et al. 1998). Because falcons are widely perceived to be among the fastest animals, substantial effort has been focused on modeling and measuring the velocities they attain in nature. Vance Tucker of Duke University has worked both with mathematical models of falcons and with live falcons to determine how fast these amazing animals can fly. By creating a mathematical model of an "ideal" peregrine falcon, he showed that peregrine falcons, *Falco peregrinus*, should be able to achieve top speeds of 89–112 m/s (that's 199–250 mph!) when diving vertically (Tucker 1998). In fact, these speeds could be even higher (138–174 m/s) if one incorporated drag into the model. What about live falcons flying in the open? Tucker and his colleagues (1998) were able to obtain flight speeds from a close relative of peregrine falcons, the 1.02 kg gyrfalcon, *Falco rusticolus*. They found that, as predicted, these birds achieved maximum speeds of 52–58 m/s during their initial (acceleratory) phase of diving. However, the speed remained relatively constant after that point and then slowly began to decline. Thus, while a top speed of 58 m/s is pretty amazing, it was substantially lower than predicted. One possible reason for this discrepancy is that birds experienced increased drag after initially accelerating, whereas the model predicted prolonged acceleration. Other data seem to verify this pattern. Peregrine falcons and goshawks both obtained impressive but presumably underachieving speeds of 19–23 m/s initially during the flapping (acceleratory) phase, followed by dives of about 30 m/s (Alerstam 1987). Because these were shallow dives, the observed speeds were similar to predicted speeds (Pennycuick 1975). However, although theoretical models suggest that there should be a positive curvilinear relationship between dive angle and speed, in fact, this relationship in the "real world" is almost level. Therefore, something has to give—either birds are modifying their behavior to allow them to do other things (e.g., hunt) or the model needs tweaking. Either way, this elegant work shows the value of combining theory with real-world observations of animals, and it shows that animals are loathe to follow our predictions!

How closely do speeds in other bird species match predicted maximum speeds? For a group of 36 bird species (Pennycuick 1997), mean speeds during undisturbed flight were 8.8–19.1 m/s; while these species clearly vary in how fast they can fly at maximum speed, these data suggest that, in general, birds are often flying at 50% of their maximum speed, at least, and, in some cases, may be moving at 70%–90% of their maximum speed. Another analysis of 48 bird species (mostly different from the above species) confirmed this trend—average speeds were 8.0–30.8 m/s when birds migrated, with most species moving at 11.0–13.0 m/s. These data show that, when birds fly, they are moving very fast, relatively speaking, compared to terrestrial animals moving when undisturbed. The exact reasons underlying this observation remain a topic of study but could relate to the unusual manner in which speed and energetic output is related in birds. In terrestrial animals, in general, energetic output increases more or less linearly with speed (at higher speeds, once anaerobic metabolism kicks in, this relationship changes); in birds, however, the relationship is believed to be more U shaped; that is, it is energetically expensive for a bird to fly either really slowly or really fast (Tobalske et al. 2003). Thus, birds may opt to fly at a reasonably fast speed that minimizes energy stores and allows them to reach their destination in a reasonable time frame.

In sum, the view for all-out effort for animals is mixed. Modern technology has allowed us to determine whether the world's premier animal athletes (e.g., falcons, cheetahs) also live up to their reputation in the wild. Both cheetahs and falcons can move at remarkable speeds; how often these animals actually engage in such high-level performance is less clear but is perhaps not as often as would have been expected. Animals often cannot afford to take unnecessary risks—if a cheetah or falcon can capture a prey item while moving far slower than usual, they likely will. Excessive speeds can injure a predator unnecessarily and, unlike the case for humans, there are no hospitals in nature for repairing broken bones! Finally, data from species such as lizards and birds show that animals often do move quite close to their maximum capacities, at least during certain behaviors.

3.6 Emergent behaviors in the field: Defying physiological constraints

Animals often exhibit emergent behaviors in nature that are not observed under more stressful laboratory contexts. It turns out that these behaviors may in some cases explain why animals can perform feats in nature that, at face value, don't seem possible. Marine mammals offer perhaps the most convincing example of this phenomenon. Since they are air breathers, like us, marine mammals must respire using the oxygen they obtain from breathing at the surface just before they dive. Based on estimates of how much oxygen a marine mammal takes in before it dives, and taking into account other factors such as the average speed of movement, it is possible to calculate how much time the animal should spend under water. This predicted duration of time is called the aerobic dive limit (ADL), and a number of studies have examined whether marine mammals can reach their ADLs. Keep in mind that any violation of the ADL could be harmful because, as aerobic metabolism becomes impossible, anaerobic metabolism kicks in, with its correlated buildup of dangerous lactic acid. Moreover, if marine mammals rely too much on anaerobic metabolism during diving, they will incur an "oxygen debt," which will require them to spend more time breathing air at the surface before diving again and thus be exposed to predators such as sharks or killer whales.

In a few species (e.g., Weddell seals, harp seals, and narwhals), the theoretical calculations of ADL are roughly correct and, in a few cases, marine mammals dive for periods that are shorter than predicted. But, for other species, notably, gray seals, elephant seals, and bottlenose dolphins, the ADL values are exceeded—sometimes by a large margin (Hindell et al. 1992; Williams et al. 2000). For example, for some bottlenose dolphins, the amount of time they spend underwater exceeds their ADL value by about 28%. The answer to this paradox may lie in the assumptions underlying how ADL values are calculated. These values generally assume constant movement from which energetic consumption can be reliably estimated. But marine mammals do not always move continuously and sometimes employ intermittent locomotion as a behavioral strategy to avoid expending energy. In intermittent locomotion, movement is punctuated by short pauses, usually lasting a second or two; thus, ATP is conserved during intermittent locomotion, relative to continuous movement. Studies with bottlenose dolphins show that these animals use a "kick-and-glide" method of ascending that alternates periods of rapid propulsion, which is generated by using the tail fin, with periods of gliding (Williams 2001). Behavior thus is a tool that allows these animals to exceed basic constraints imposed by their physiology.

3.7 Slackers and overachievers: Behavioral compensation in nature

Remember the lyrics from the Rolling Stones, "you can't always get what you want, but sometimes you get what you need"? That pretty much sums up the way to think about animal performance in the real world. Rather than expecting animals to perform at 100% of maximum every time they have the opportunity, rather think of them as doing only what they have to do to survive.

This view leads directly to the concept of behavioral compensation, that is, the process of performing certain actions to compensate for a particular weakness. In the case of animal performance, the consequence of behavioral compensation is that the functionally limited individual performing such compensation must exert increased effort, as compared to a normally functioning individual. The flip side of the coin is that animals that are not functionally limited are not exerting 100% of their effort. Juveniles are a case in point. The juveniles of most species are small and usually slower relative to mature adults. This fact, along with their smaller amount of experience, makes them especially vulnerable to predators, many of which focus their energies on capturing them and ignoring adults. Behavioral compensation seems to occur during escape locomotion in several kinds of lizard species, as "slower" juveniles will run close to their maximum speed, while "faster" adults will not work as hard (Dial et al. 2008). Similarly, different species of *Anolis* lizards show the same pattern. "Slow" species run at close to 100% of their maximum capacities, whereas "fast" species do not (fig. 3.9). Another way to think about this pattern is in terms of "slackers" and "overachievers" (Irschick et al. 2005). Larger, inherently fast lizards (i.e., adult male lizards) can be viewed as "slackers"; they will run hard in the laboratory but achieve far less than their maximum capacities in the field. On the other hand, small, inherently slow lizards (i.e., juvenile lizards) can be viewed as "overachievers"; they run more slowly in laboratory settings compared to larger lizards but will run at nearly 100% of maximum capacity in their natural environment. Within birds, there is a similar negative correlation

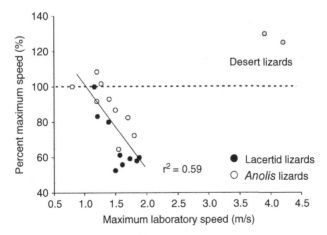

Fig. 3.9 A plot of maximum sprint speed as measured in the laboratory versus percent maximum speed as used by different lizard species and age classes in the field in response to a potential predator. Note the negative relationship between the two variables (with the exception of two desert lizards in the upper right); this result indicates that "fast" lizard species and age classes (slackers) exert far less "effort" than "slow" lizard species and age classes (overachievers) do. Taken from Irschick et al. (2005) with permission from Wiley-Blackwell press.

between total power capacity and percent maximum power used (Dial et al. 2008);this observation suggests that power-limited species tend to use all, or at least most, of their power capacities, whereas nonlimited species seem to use far less.

References

Alerstam T. 1987. Radar observations of the stoop of the peregrine falcon *Falco peregrinus* and the goshawk *Accipiter gentilis*. Ibis 129:267–273.

Andrews RD, Jones DR, Williams JD, Thorson PH, Oliver GW, Costa DP, Le Boeuf BJ. 1997. Heart rates of northern elephant seals diving at sea and resting on the beach. Journal of Experimental Biology 200:2083–2095.

Blake RW, Kolotylo R, de la Cueva H. 1990. Flight speeds of the barn swallow, *Hirundorustica*. Canadian Journal of Zoology 68:1–5.

Davis RW, Polasek L, Watson R, Fuson A, Williams TM, Kanatous SB. 2004. The diving paradox: New insights into the role of the dive response in air-breathing vertebrates. Comparative Biochemistry and Physiology Part A 138:263–268.

Dial KP, Green E, Irschick DJ. 2008. Allometry of behavior. Trends in Ecology and Evolution 23:394–401.

Dudley R, Srygley RB. 1994. Flight physiology of neotropical butterflies: Allometry of airspeeds during natural free flight. Journal of Experimental Biology 191:125–139.

Fossette S, Schofield G, Lilley KSM, Gleiss AC, Hays GC. 2012. Acceleration data reveal the energy management strategy of a marine ectotherm during reproduction. Functional Ecology 26:324–333.

Gleiss AC, Norman B, Wilson RP. 2011. Moved by that sinking feeling: Variable diving geometry underlies movement strategies in whale sharks. Functional Ecology 25:595–607.

Hammerschlag, N, Luo J, Irschick DJ, Ault JS. 2012 A comparison of spatial and movement patterns between sympatric predators: Bull sharks (*Carcharhinus leucas*) and Atlantic tarpon (*Megalops atlanticus*). PLoS ONE. 10.1371/journal.pone.0045958.

Hedenström A, Alerstam T. 1992. Climbing performance of migrating birds as a basis for estimating limits for fuel-carrying capacity and muscle work. Journal of Experimental Biology 164:19–38.

Higham TE, Stewart WJ, Wainwright PC. 2015. Turbulence, temperature, and turbidity: The ecomechanics of predator-prey interactions in fishes. Integrative and Comparative Biology 55:6–20.

Hindell MA, Slip DJ, Burton HR, Bryden MM. 1992. Physiological implications of continuous, prolonged, and deep dives of the southern elephant seal (*Mirounga leonine*). Canadian Journal of Zoology 70:370–377.

Irschick DJ, Garland T Jr. 2001. Integrating function and ecology in studies of adaptation: Investigations of locomotor capacity as a model system. Annual Reviews of Ecology and Systematics 32:367–396.

Irschick DJ, Herrel A, Vanhooydonck B, Huyghe K, Van Damme R. 2005. Locomotor compensation creates a mismatch between laboratory and field estimates of escape speed in lizards: A cautionary tale for performance to fitness studies. Evolution 59:1579–1587.

Irschick DJ, Jayne BC. 1999. A field study of effects of incline on the escape locomotion of a bipedal lizard, *Callisaurus draconodies*. Physiological and Biochemical Zoology 72:44–56.

Irschick DJ, Losos JB. 1998. A comparative analysis of the ecological significance of locomotor performance in Caribbean *Anolis* lizards. Evolution 52:219–226.

Mandel JT, Bildstein KL, Bohrer G, Winkler DW. 2008. Movement ecology of migration in turkey vultures. Proceedings of the National Academy of Sciences 105:19102–19107.

McLaughlin RL, Montgomerie RD. 1985. Flight speeds of central place foragers: Female Lapland longspurs feeding nestlings. Animal Behaviour 33:810–816.

Ovaskainena O, Smith JL, Osborne DR, Carreck NL, Martin AP, Niitepo K, Hanski I. 2008. Tracking butterfly movements with harmonic radar reveals an effect of population age on movement distance. Proceedings of the National Academy of Sciences 105:19090–19095.

Pennycuick CJ. 1975. Mechanics of flight. In Farner DS, King JR, Parkes KC, eds. Avian Biology, Vol. 5. Academic Press, New York, NY, pp. 1–75.

Pennycuick CJ. 1997. Actual and "optimum" flight speeds: Field data reassessed. Journal of Experimental Biology 200:2355–2361.

Pennycuick CJ, Åkesson S, Hedenström A. 2013. Air speeds of migrating birds observed by ornithodolite and compared with predictions from flight theory. Journal of the Royal Society Interface 10:20130419.

Tobalske BW, Hedrick TL, Dial KP, Biewener AA. 2003. Comparative power curves in bird flight. Nature 421:363–366.

Tucker VA. 1998. Gliding flight: Speed and acceleration of ideal falcons during diving and pull out. Journal of Experimental Biology 201:403–414.

Tucker VA, Cade TJ, Tucker AE. 1998. Diving speeds and angles of a gyrfalcon (*Falco rusticolus*). Journal of Experimental Biology 201:2061–2070.

Turchin P. 1998. Quantitative Analysis of Movement: Measuring and Modeling Population Redistribution in Animals and Plants. Sinauer Press, Sunderland, MA.

Wakeling JM, Hodgson J. 1992. Optimisation of flight speed of the little, common and sandwich tern. Journal of Experimental Biology 169:261–266.

Webb PW. 2006. Stability and maneuverability. In Shadwick RE, Lauder GV, eds. Fish Biomechanics, Vol. 23. Academic Press, New York, NY, pp. 281–332.

Welham CVJ. 1994. Flight speeds of migrating birds: A test of maximum range speed predictions from three aerodynamic equations. Behavioral Ecology 5:1–8.

Williams TM. 2001. Intermittent swimming by mammals: A strategy for increasing energetic efficiency during diving. American Zoologist 41:166–176.

Williams TM, Davis RW, Fuiman LA, Francis J, LeBoeuf BJ, Horning M, Calambokidis J, Croll DA. 2000. Sink or swim: Strategies for cost-efficient diving by marine mammals. Science 288:133–136.

Wilson AM, Lowe JC, Roskilly K, Hudson PE, Golabek KA, McNutt JW. 2013. Locomotion dynamics of hunting in wild cheetahs. Nature 498:185–189.

4 | The ecology of performance III
Physiological ecology

4.1 Environmental influences on performance

The natural environment and the performance capacities of animals exhibit a complex relationship. The environment clearly impacts how animals can perform on a day-to-day basis and how performance capacity evolves over the span of many years. Some examples of environmental variables that have clearly influenced the evolution of animal performance include temperature, ambient air or water flow conditions, the density of substrate in the environment in which animals live (e.g., water, soil), and the amount of oxygen available. The study of how such factors influence animal physiology and performance falls into the realm of physiological ecology. For example, temperature influences most physiological processes, and low or high temperatures can impair movement because of their negative effects on muscle function. Because natural habitats vary in temperature, animals must evolve anatomical or behavioral adaptations for coping with such extremes. Consequently, animals exhibit a range of physiological thermal adaptations that allow them to tolerate a wide range of temperatures, including some pretty spectacular ones! Some desert ants can tolerate ambient air temperatures of 50°C (122°F), and some fish have "antifreeze" elements in their blood, thus enabling them to swim in arctic waters! However, many animals are not capable of evolving such thermal tolerances and are thus more limited in the kinds of habitats they can live in. Environmental variation therefore both acts as a constraining factor and as an innovating factor on animal physiology and function. It also dictates, to some extent, when animals are active.

In addition to studying how the environment influences animal performance, another aim of physiological ecology is to understand how the

Animal Athletes. Duncan J. Irschick and Timothy E. Higham. © Duncan J. Irschick and Timothy E. Higham 2016. Published in 2016 by Oxford University Press.

physiology of animals interfaces with their environment; it therefore includes the study of how animals obtain and expend energy or consume and use oxygen, among other examples. To understand how the environment influences animal performance and how animals use the environment, it is useful to take both laboratory-based and more naturalistic approaches. This integrative approach is important because, whereas laboratory studies can reveal the mechanisms for tolerating and using the environment, only field studies can reveal the actual environmental challenges that animals encounter. Moreover, as seen in this chapter, animals exhibit a range of behaviors that complement their anatomical specializations and thereby enhance their ability to tolerate or even manipulate the environment. Given the vast complexity of the environmental challenges that animals face, we will focus on two aspects that have had pervasive influences on animal performance: temperature and energetics.

As a point of definition, by physiology we mean the internal processes that govern the ability of animals to perform basic metabolic tasks, such as respiration, balancing ions with water, and using their muscles, or other powering structures, to propel movement. Physiology is composed of many hierarchical layers, including individual cells, tissues, and organs, and the properties at each of these levels limits the ability of the whole body, or parts of the body, to perform dynamic movements.

4.2 Energetics and movement

Movement requires energy. Because movement is necessary for many animals, they have evolved evolutionary specializations, both morphologically and physiologically, to minimize this cost. How much energy is expended during various kinds of movement varies according to body size, life history, gait, and many other aspects. One can examine the topic of energetic cost in many ways, but let's first define this term and then examine how it changes with three factors: animal size, the speed and type of movement, and the kinds of gaits that animals use. These three factors have generally been studied for steady-state locomotion, mostly in mammals, and typically on some kind of treadmill or other similar substrate.

Measurements of energetic cost are usually taken by quantifying how much oxygen an animal uses during an arduous task, such as running steadily on a treadmill. While it would be tempting to also perform such experiments when animals are accelerating over short distances, this latter form of movement is typically fueled by anaerobic metabolism, which does not require oxygen. Energetic cost is the amount of oxygen used during movement relative to the total mass of the animal. This phenomenon is expressed as the cost of locomotion and is equivalent to "gas mileage" in cars, as it describes how much energy an animal expends as it moves a certain distance.

Body size has perhaps the most profound effect on the energetics of locomotion. Body size influences energetics because of the geometric manner by which animals grow. As animals grow larger in body size, almost all other aspects grow as well, including the total mass of muscle, and the length and mass of limbs that propel movements. The pattern by which animals change in size, either within or among species, is called scaling and is usually examined at two levels: within species (e.g., during ontogeny) and among species (e.g., evolutionary). Mammals provide a wonderful system in which to examine the influences of body size on locomotor energetics because of their vast range in size, as elephants are about a million times more massive than shrews. Counterintuitively, it is far more energetically expensive to be a shrew than an elephant, as basal (or resting) metabolic rate decreases dramatically as mammals become larger. Once mammals are measured in a common locomotor task, this trend does not change. The energetic cost of locomotion is far higher for small mammals than for large ones (Tucker 1970). These two trends may explain in part why longevity in mammals increases with body size, given the limited energy budgets in most organisms. However, this same pattern of decreasing cost of transport as animal size increases also holds true for many other kinds of animals, including insects, amphibians, reptiles, and mammals. The similarity of energetic costs for locomotion among animals of widely varying shapes and life styles is remarkable and suggests that there are common mechanisms by which energy is expended during locomotion.

A second factor is the speed of locomotion. Anyone who has been walking somewhere only to realize that they were late and then had to start running knows that they had to work a lot harder when running! This supposition is borne out with experimental data showing that, for a variety of animals, the rate of oxygen consumption increases with running speed (Heglund et al. 1982; Taylor et al. 1982). Note that the rate of oxygen consumption is not the same as the cost of transport discussed above, because the latter measure takes into account energetic cost per unit distance, whereas the former does not. How rapidly oxygen consumption rises with locomotor speed is size dependent, at least within some animal groups. For example, within mammals, smaller species experience a more rapid increase in the rate of oxygen consumption, as they run faster compared to larger mammals. The general rule of increased oxygen consumption with increased speed does not hold for every animal. One exception are birds, for which the rate of oxygen consumption follows a more "U-shaped" pattern with speed, with high oxygen consumption rates occurring at low and high speeds, and low values at intermediate speeds (Tobalske et al. 2003). When bird flight is examined in terms of cost of transport, birds display poor "gas mileage" when flying very slowly or fast, and far better gas mileage when flying at intermediate speeds, which can still be substantial (5–8 m/s). Other exceptions are kangaroos and their close relatives. In these species, the rate of oxygen consumption increases rapidly with speed when they are moving slowly using all four limbs plus their tail (termed pentapedal gait). However, as the animal hops on their large hindlimbs and gathers speed, its rate of oxygen consumption does not change appreciably owing to the utilization of its large and stiff tendons as giant springs to efficiently store and release elastic energy with each hop (Dawson and Taylor 1973).

The speed of locomotion in animals is often correlated with different gaits, which are coordinated sets of limb cycles, such as the trot or the gallop. While gaits have been known for many years, it was not until the 1980s that their functions were elucidated through a series of studies performed at Harvard University by Charles Taylor and his colleagues (Heglund et al. 1982; Taylor et al. 1982). All mammals use gaits in some fashion. If you simply compare the movements of your cat or dog when they are walking to those when they are running, you will notice the use of galloping at higher speeds. Among animals, horses have been the best studied in terms of their gaits. Although horses have been purported to have many gaits, three of the best known are the walk, the trot, and the gallop, with the "walk-trot" or "trot-gallop" transitions between them. These gaits in horses act much like gear transitions in cars, such as when one shifts from first to second gear. As horses reach the speed limit of each gait, the energetic cost increases, much like when an engine revs at high and unsustainable rates when moving too fast for a given gear (Hoyt and Taylor 1981). By "shifting" to a new gait, such as from walking to trotting, the energetic cost then decreases, in the same way that a car's engine revs less rapidly when shifting. In short, gaits seem to save energy during movement by allowing the animal to move in a more efficient manner. Whether gaits in different animals all follow this rule would benefit from further study.

Finally, anyone who has struggled to swim the many laps imposed on them by a vigilant and seemingly malicious swim coach would surely claim that swimming is more energetically expensive than walking or running. Older work (Schmidt-Nielsen 1972) found that the cost of transport was highest for walking and running animals, lowest for swimming animals, and intermediate for flying animals. However, that comparison was among disparate taxonomic groups, and once different mammals were compared for these forms of locomotion, the cost of transport did not differ among walking/running, swimming, and flying (Williams 1999). This counterintuitive pattern might be explainable by the kinds of specializations discussed in this book. For example, most fish or marine mammals possess streamlined bodies with flattened fins that they use to propel themselves, and the apparent benefit of these traits seems evident from the low energetic costs of their movements in media such as water.

4.3 Aerobic versus anaerobic metabolism

Variation in performance capacity, both among individuals and among species, often arises because of differences in physiology. Anyone who has watched the human Olympics has probably noticed the division of long-distance and short-distance events, and the corresponding differences in body shape and style of the athletes in these different events. The ability to run long distances is typically dictated by aerobic metabolism, whereas the ability to run short distances, such as a 100 m dash, is driven by anaerobic metabolism. Most animals and humans utilize both processes simultaneously to some degree; for example, during movements of intermediate intensity, both processes may be recruited to fuel movement.

Among and within animal species, there are vast differences in aerobic and anaerobic capacities. If one takes the time to place a spider on a large sheet of paper and gently prod it with a pencil, the terrified spider will likely flee but within a few minutes will become exhausted. By comparison, anyone who has ridden a fit and well-trained horse knows that horses can gallop at high speeds for over a mile before becoming fatigued. There are several factors that explain this difference; but first, some terminology is required. Aerobic metabolism is used when animals consume oxygen and produce carbon dioxide and is driven by the mitochondria in cells. The process of aerobic respiration is relatively efficient compared to anaerobic metabolism, as it produces relatively large amounts of energy (ATP) per cycle. More importantly, aerobic metabolism is more sustainable than anaerobic metabolism. Whereas aerobic metabolism produces carbon dioxide as a waste product that can be removed from the body via respiration, anaerobic respiration produces lactic acid, which is commonly thought to be deleterious to powering muscles and the accumulation of which results in that nasty (or pleasant, depending on your point of view) "burn" in our muscle after an intense workout.

In some cases, excessive lactic acid buildup following intense exertion can be deadly for animals. However, recent research has raised questions about whether the buildup of lactic acid is in fact harmful at physiological levels. Strenuous exercise results in an increase in extracellular potassium, thus decreasing both muscle excitability and force production. Surprisingly, lactic acid buildup, and the consequent acidosis, can counteract the effects of elevated extracellular potassium and actually prevent muscle fatigue (Nielsen et al. 2001)! This observation provides a great example of why correlation studies should be interpreted with caution, as lactic acid buildup was thought to produce muscle fatigue because, historically, it was found that lactic acid levels increased when fatigue (decreased muscle force) increased. However, the finding that lactic acid can prevent muscle fatigue doesn't obviate the fact that the discomfort associated with an increased level of acid in your muscles might have an impact on your desire to exercise! Anaerobic metabolism is relatively inefficient, at least over longer time periods of movement, as it produces less ATP per cycle than aerobic metabolism does, although some of the waste products from lactate are converted back into an alternative form of energy production. Further, anaerobic metabolism does not require oxygen as input and thus offers the benefit that it is not limited by the ability of organisms to ventilate oxygen into the cells.

However, the above examples of the horse and spider seem to run counter to these principles. If anaerobic metabolism is unsustainable, why can horses gallop for long distances? The answer lies in a common metric used to quantify aerobic endurance, termed maximum VO_2, or VO_{2max} (also referred to as "aerobic capacity"). VO_{2max} is the maximal energy used to move that can be generated by oxidative phosphorylation in mitochondria (Weibel et al. 2004) and thus is a good metric of aerobic capacity. One of the added benefits of a high value of VO_{2max} is the ability to run at relatively high speeds using aerobic as opposed to anaerobic metabolism (fig. 4.1); however, as speed increases,

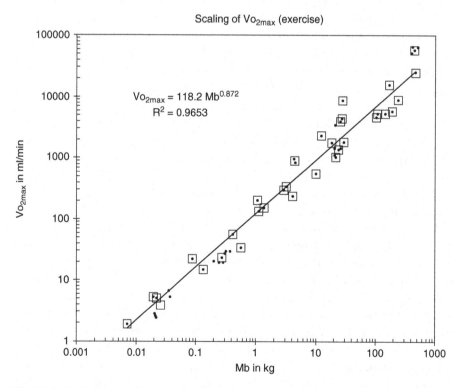

Fig. 4.1 A plot showing the linear relationship between body size (Mb) and VO_{2max} among 34 mammal species. Note that, as mammals become larger, their VO_{2max} values increase. Taken from Weibel et al. (2004) with permission from Elsevier Press.

anaerobic metabolism is gradually recruited until the speed at which VO_{2max} is reached, after which it is heavily recruited. There is notable variation in VO_{2max} among animal species, and mammals alone vary in VO_{2max} by more than a thousandfold! For example, horses have a relatively high VO_{2max}; thus, they can run at relatively fast speeds and still run aerobically. To some extent, VO_{2max} can be altered by exercise. Human athletes that train for marathons can improve their VO_{2max}, but this increase will disappear soon after exercise ceases and does not compare in size to the natural variation among animal species.

4.4 Born to run: Variation among animal species in the energetics of movement

The variation in VO_{2max} among species, and the other life-history characteristics that have coevolved with this trait, can be appreciated by comparing mammals and reptiles. In general, most mammals have high values of VO_{2max} compared to reptiles (Table 4.1; Bennett and Ruben 1979; Bennett 1991). By contrast, reptiles typically have low values of VO_{2max} and quickly resort to anaerobic metabolism as they begin to move quickly (Bennett 1991). This physiological divide has profound consequences for the ecology and life history of these groups. Whereas mammals can move widely and employ hunting or foraging strategies that require aerobic abilities, reptiles are more limited and typically rely on ambushing strategies with quick and short movements. Two factors may explain this physiological divide. First, mammals are endothermic, whereas reptiles are ectothermic. By being endothermic, mammals can maintain a relatively high and constant internal temperature that enables muscles to operate in a more consistent and efficient fashion. Second, mammals seem to have a more efficient system for acquiring oxygen and removing carbon dioxide than reptiles do. The mammalian four-chambered heart, large lung volumes, and upright gait enable high ventilation rates and an efficient cardiovascular pulmonary system. Reptiles have three-chambered hearts and more sprawling gaits. Indeed, there is some evidence that the lateral flexion of the body during locomotion in lizards may limit their ability to intake oxygen because of constraints on lung volume (Carrier 1987). However, some reptiles seem to defy their reputation as aerobic wimps. Varanid lizards, otherwise known as monitor lizards, are large and generally active lizards (the Komodo dragon is the largest member of this group) that can not only run fast but also run for relatively long periods. One reason for this impressive aerobic performance is the presence of a gular pump in some species (Owerkowicz et al. 1999). The gular pump acts like a bellows by assisting in delivering oxygen to the lungs (fig. 4.2), thereby generating a higher rate of oxygen exchange and improved aerobic capacity. Other modifications, such as a more erect gait,

Table 4.1. A table showing the production of ATP over five minutes of intense activity in rodents and reptiles. Note the much greater contribution of aerobic metabolism for mammals, and the greater contribution of anaerobic metabolism for reptiles. Table redrawn from Bennett and Ruben (1979). The greyhound image is from Wikimedia Commons and the lizard image is from Duncan J. Irschick.

Species	Total ATP (μmole/g)	ATP: Aerobic (μmole/g)	ATP: Anaerobic (μmole/g)	Body mass (g)
Montane vole, *Microtus montanus*	53	40	13	25
Meriam's kangaroo rat, *Dipodomys merriami*	107	98	9	35
Western fence lizard, *Sceloporus occidentalis*	46	24	22	13
Coachwhip snake, *Masticophis flagellum*, and racer snake, *Coluber constrictor*	52	23	29	262

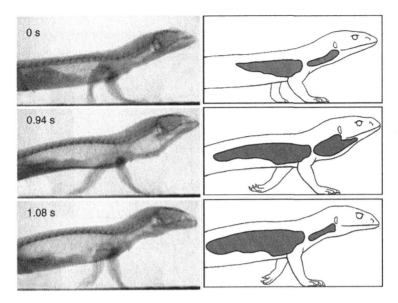

Fig. 4.2 X-ray negative images (on the left) showing locomotion in the monitor lizard *Varanus exanthematicus*, with panels on the right showing the different stages of a breath cycle. These are (top to bottom): end of exhalation, end of gular and costal inspiration, and at the end of the gular pump. These lizards employ the gular pump like a bellows that funnels air into their lungs, thereby improving overall aerobic performance relative to other lizards. Redrawn from Owerkowicz et al. (1999).

a functionally divided heart, and an enlarged lung surface area that facilitates gas exchange, could all explain the improved stamina of these lizards relative to other reptiles. This example shows that, while physiological and anatomical constraints exist, the origin of evolutionary novelties can dramatically enhance performance as well.

The evolution of high levels of VO_{2max} has occurred within mammals as well to meet the demands of their distinctive habitats. For example, the VO_{2max} for the pronghorn, *Antilocapra americana*, has a notably high value relative to that of other mammals and may have evolved because of the need for pronghorns to run at fast speeds for long distances when fleeing fleet predators such as cheetahs, which can run quickly but only for short time periods. The demands of flight are also met by high aerobic capacities, as evidenced by the high VO_{2max} values found for hummingbirds and bats.

While locomotion is generally energetically expensive, the need for animals to move long distances presents the challenge of moving long distances at low energetic costs. One method to reduce the total energetic cost of locomotion is to move intermittently, such as when animals "pause" for very short periods, sometimes for even less than a second (Gleeson and Hancock 2001).

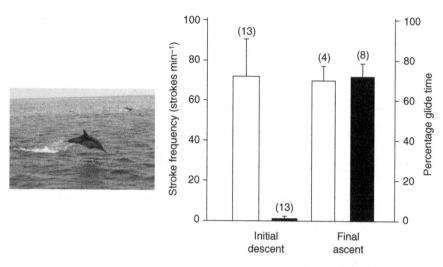

Fig. 4.3 *Left*, Image of a dolphin in the wild. *Right*, Stroke frequency and percent time gliding for dolphins swimming freely in nature. Note the greater use of the "kick-and-glide" method on the final portion of the ascent, when oxygen stores are at their lowest. This use of intermittent locomotion likely allows these animals to spend less energy when diving. Graph taken from Williams et al. (1999) with permission from the Company of Biologists Ltd. Dolphin image is from Wikimedia Commons.

Intermittent locomotion can appear "herky-jerky" or even spasmodic because of the short pauses intermixed with bouts of movement. A good analogy is soccer—players run quickly for short periods, followed by several seconds or more of rest, followed by another bout of running. During these pauses, the body recovers some energy in the form of ATP. Intermittent locomotion therefore allows animals to move long distances at a lower energetic cost relative to unbroken movement, with the only cost being time. For instance, in the desert gecko *Teratoscincus przewalskii*, intermittent locomotion results in about a 1.7-fold increase in the total distance moved, relative to continuous movement (Weinstein and Full 1999). This mode of locomotion is widely used by many kinds of animals, but some documented examples of animals that use intermittent locomotion are dolphins and lizards, among others. Dolphins, for example, intermix periods of "kicking" with gliding (Williams et al. 1999; Williams 2001), especially upon ascent, when their oxygen stores are at their lowest in a dive, and therefore can dramatically reduce their total need for oxygen to move a certain distance (fig. 4.3).

4.5 Feel the burn: Harmful byproducts of movement

In addition to energy being spent as a cost for movement, there are additional physiological costs to some kinds of locomotor performance or vigorous activity. You have probably heard the phrase "feel the burn," which refers to the painful burning sensation from intense physical activity such as weight-lifting or sprinting. The "burn" is caused by the buildup of lactic acid in muscles and other tissues in the body and is a signal that muscles are rapidly fatiguing and will soon be exhausted. Lactic acid is produced as a byproduct of anaerobic metabolism, which is generally used during intense movements, such as when sprinting at maximum capacity, or lifting large weights. The muscles used to drive such movements are typically fast-twitch muscles that produce large amounts of force but that also fatigue easily. However, anaerobic metabolism is also used by some animals as a normal way to acquire energy when oxygen levels are low, such as when animals dive for long periods underwater. For these animals, the production of lactic acid could be dangerous; if an animal stays under too long, lactic acid levels could build to high levels, potentially weakening the muscles and making the animals more susceptible to predators, for example (however, see Section 4.3). So lactic acid buildup occurs in two ways: as a result of intense exertion and as a result of moving in an environment without sufficient oxygen. The challenge of lactic acid buildup is especially problematic because, even after intense exercise has stopped, levels continue to build within the body, and elevated levels of lactic acid can persist for tens of minutes; thus, very intense exercise can both impair an animal and devour large amounts of oxygen. This process is termed excess postexercise oxygen consumption, or EPOC (Gleeson and Hancock 2002). Diving animals have been well studied in terms of their lactic acid production following long dives, and there is some evidence that they can dive for very long periods without incurring a substantial lactic acid debt, despite theoretical predictions that they should incur one. The primary reason for reduced lactic acid level buildup seems to be the use of intermittent locomotion when ascending, which conserves ATP and reduces muscle fatigue.

Despite these tricks, the costs of diving nevertheless occur and, for most animals, the use of intermittent locomotion amounts to damage control. For some mammals and birds, the effects of prolonged diving are somewhat ameliorated by a metabolic subdivision in which their core and periphery become anaerobic, whereas their brains remain aerobic. Maintaining the aerobic capacity of their brains is important because of their high demand for oxygen, and brain impairment would mean reduced decision-making ability in a tough spot. However, in some diving turtles, both the brain and the entire body become anaerobic, resulting in extremely high lactic acid levels, a problem that is somewhat alleviated by the buffering ability of the shell and bones (Robin et al. 1979). Another ameliorating factor is that, in turtles, as well as in most animals that face anoxic conditions, the ability to depress metabolic rate lowers the overall energy demands, which in some cases can result in animals in deep areas moving in an almost zombie-like state. When metabolic rates are extremely low, the body has little requirement for energy through either aerobic or anaerobic metabolism. For example, the average heart rate of elephant seals near the surface is around 100 beats/min but is reduced to around 4 beats/min on very deep dives (Andrews et al. 1997)! The ability to dampen metabolic rate may explain why some animals, such as sea snakes, seem to build up relatively little lactic acid during prolonged dives.

4.6 Temperature and animal performance

We have both spent a good portion of our scientific lives working in deserts, including the Mojave Desert of North America and the Namib Desert of Africa, both of which exhibit ambient temperatures ranging from 0°C to 45°C (Namib) and 0°C to 38°C (Mojave) between the summer and winter. Moreover, over the course of a single day, ambient temperatures can range from 10°C at night to over 40°C during the day. Surface temperatures, driven largely by radiant heat, can reach over 50°C, which is far higher than the maximum tolerable temperature for most ectotherms. In short, any desert animal must display some ability to tolerate both cold and hot temperatures across different seasons. It also means that any desert animal cannot simply afford to move passively and incautiously in such an extreme thermal habitat. Although the temperatures found in these desert environments are extreme, temperatures can range significantly in many other environments as well; thus, the ability to respond to changes in temperature has shaped much of the thermal and behavioral diversity of animals.

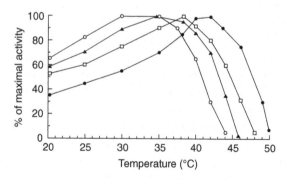

Fig. 4.4 A typical temperature-performance curve showing relatively poor performance and a long tail at lower temperatures, optimal temperatures over a plateau, and then a precipitous decline at higher temperatures. Data are for adenosine triphosphatase (ATPase) from four species of lizards: *Dipsosaurus dorsalis* (solid circles), *Gerrhonotus multicarinatus* (open circles), *Sceloporus undulatus* (triangles), and *Uma notata* (squares). Redrawn from Angilletta (2009), and data originally from Licht (1967) with permission from Oxford University Press.

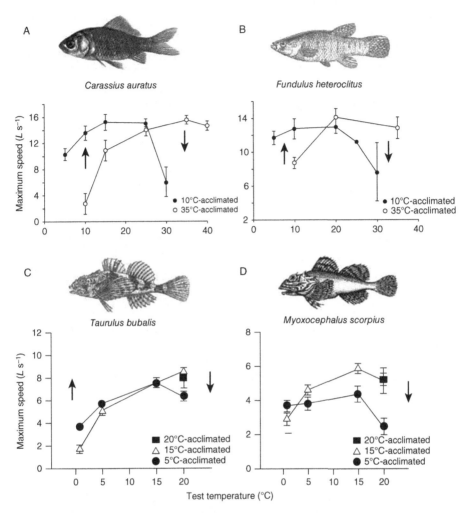

Fig. 4.5 Data on thermal plasticity of burst-speed locomotion in four fish species, from Johnston and Temple (2002). Data shown are for maximum length-specific speeds during escape. Fish were acclimated to different temperatures, as noted in the graphs. The direction of the acclimation response for a given temperature is noted by arrows, whereas a horizontal line indicates no change. Taken from Johnston and Temple (2002) with permission from the Company of Biologists Ltd.

Of all environmental variables that influence performance capacity, temperature has been the best studied and is perhaps the most ecologically relevant, especially for ectotherms. Temperature plays an overarching role in nearly every aspect of performance, and its effects have been extensively studied in many kinds of ectotherms (and also some endotherms), from tiny amoebas to large pelagic fish. The simple and profound influence of temperature occurs for several reasons. First, nearly all enzymes that enable muscles to function are temperature sensitive and are most efficient within a relatively narrow range of potential temperatures, usually exhibiting a slow decline in activity as temperature declines, and a precipitous decline at very high temperatures (fig. 4.4). If enzymes and other molecules cannot effectively function at very high or low temperatures, then the higher-level tissues that these enzymes influence also cannot function effectively, resulting in suboptimal performance, as exemplified in a temperature-performance curve for maximum burst speed in fish (fig. 4.5). This curve means that most, but not all, performance traits will be suboptimal at lower temperatures and will slowly improve, with increasing temperature, to a peak. There is typically a broad plateau of temperatures at which performance is maximal or near-maximal, and then there is usually a precipitous decline at temperatures above the plateau; this observation underscores the fact that, whereas cold temperatures may diminish performance, high temperatures can easily kill organisms. This asymmetric effect of temperature on performance has stark consequences for ectotherms in particular, who cannot regulate their internal body temperature. Nonetheless, the influence of temperature on performance is not negligible for endotherms, which can also suffer poor performance in response to very high or low temperatures. One of the best-studied performance traits in relation to temperature is the speed of locomotion. Locomotor speed can be measured in different ways. One way is to measure it over relatively long distances, such as several meters for smaller animals; in this case, it is typically referred to as sprint speed. Alternatively, one can measure maximum speed over much shorter distances, such as over centimeters or millimeters; in this case, it is called burst speed. This second measure is particularly appropriate when examining small aquatic organisms that rely on rapid bursts of locomotion to elude predators. Across a variety of ectotherms, ranging from fish to snakes, lower body temperatures results in greatly diminished sprint speed and burst speed. Further, reduced body temperatures also dramatically reduce maximum acceleration, as well as endurance in many animals.

Much of the effects of temperature on locomotor performance can be understood by examining thermal effects on muscular properties that drive movement. Just as most aspects of performance show an asymmetrical bell-shaped curve when related to temperature, the ability of the limb to cycle—termed stride frequency for terrestrial animals, and stroke frequency for aquatic animals—shows a similar pattern. However, temperature does not affect the properties of muscles equally; in general, temperature has a greater effect on how fast muscle fibers contract than on the amount of power they produce. This means that, for high-power movements that are relatively slow, temperature is likely to have little effect whereas, for low-power high-frequency movements, temperature is likely to have a stronger effect. Locomotion in slow-moving and large tortoises is a case study in which variation in temperature is likely to have less influence, as their limb movements are forceful but slow (Zani et al. 2005). Another example is clinging ability in pad-bearing lizards, such as geckos. The adhesive force from the toepads likely derives from van der Waals forces, and there is little strong evidence that clinging ability is influenced by temperature (Bergmann and Irschick 2005). This is one of the intriguing features of lizard toepads, as their toepads work well in cold or hot climates!

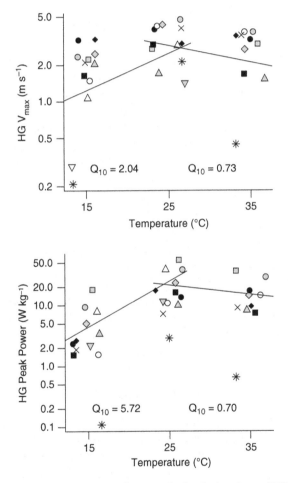

Fig. 4.6 Plots of temperature versus contractile properties for the hyoglossus (HG) tongue muscle in chameleons. Note that both variables show a change with temperature across lower and intermediate temperatures (14°C–26°C), but there is no temperature effect between 24°C and 36°C; HGV_{max} is the peak velocity for the tongue muscle, whereas HG peak power is the estimated peak power from the tongue muscle during projection. Taken from Anderson and Deban (2012) with permission from the Company of Biologists Ltd.

One way of quantifying the effects of temperature is using a value called Q_{10}. This value refers to how much of a change in a trait occurs with a 10°C shift in temperature. So, a Q_{10} of 2 for sprint speed would mean that a doubling of speed from 10°C to 20°C, for example. In tree frogs, *Hyla crucifer*, the Q_{10} values show how variation in temperature affects overall movement speed more than the amplitude of body parts, as the Q_{10} for amplitude is only 1.7, whereas for velocity it is over 5 (John-Alder et al. 1989). The implication of this trend is that animals that rely on rapid contractions more than long strides are more likely to suffer poor performance at low temperatures. The role of the muscles in these frogs also reveals some other general patterns. Some muscle properties, such as isotonic shortening, are somewhat more sensitive to temperature than force generation is. When these frogs move at very low temperatures, such as −8°C, the stroke frequency of the limbs is limited by twitch contraction time, and thus they cannot move their limbs quickly.

From our discussion above, it is clear that temperature strongly impacts the function of muscles. However, not all motions are reliant on fast muscle contraction. For example, many animals that execute ballistic or high-powered movements do so by storing elastic energy in tendons. Tongue projection is a case in point. Chameleons are slow vertebrates that zero in on prey and fling their tongue out at remarkable speeds. This incredibly fast movement can reach accelerations of 41 *g*, an incredible feat for any animal! Chris Anderson, who was a Ph.D. student at the University of South Florida at the time, examined how changes in temperature (over a range of 20°C) impacted the performance of tongue projection in chameleons (Anderson and Deban 2012; see fig. 4.6). Surprisingly, performance was invariable over the more ecologically relevant temperature range (24°C–36°C). This result indicates that the elastic recoil mechanism circumvents the constraints that low temperatures impose on muscle properties, likely via the storage of elastic energy in tendons and other elastic tissues in the feeding apparatus. This is a valuable trick for chameleons, as life persists, even when temperatures drop!

4.7 Acclimation and temperature

In addition to the ballistic methods noted above, many animal species possess a remarkable ability to acclimate to suboptimal temperatures and therefore improve their performance capacities in suboptimal temperatures. Acclimation to suboptimal temperatures is well documented in animal species, such as fish that live in shallow water or in habitats that are prone to drought, that are exposed to a variety of temperatures. Both goldfish (genus *Carassius*) and killifish are known for their ability to acclimate to different temperature regimes, a useful ability given the neglect that goldfish often undergo at the hands of their owners! If one allows these two species to acclimate to a low temperature, such as 10°C, they both increase the activity of an enzyme called fast muscle myofibrillar ATPase, as compared to when they undergo acclimation at 35°C (Johnson and Bennett 1995; Johnston and Temple 2002). Acclimation to low temperatures in these species also improved muscle twitch contraction time, which is a key trait that declines at low temperatures, therefore directly impacting locomotion. In sum, acclimation enabled goldfish to improve their escape locomotor speed at low temperatures by about 6.8-fold relative to unacclimated locomotion, whereas killifish showed a more modest 50% improvement in locomotor performance; these results showing how some species are better acclimators than others.

Laboratory studies of locomotion and muscle function show that carp are especially proficient at acclimating to a variety of temperatures. One advantage they possess is the ability to move rapidly by employing slow-twitch aerobic muscle, rather than the more commonly used fast-twitch anaerobic muscle (Rome et al. 1990). Changes in the biochemistry and morphology of

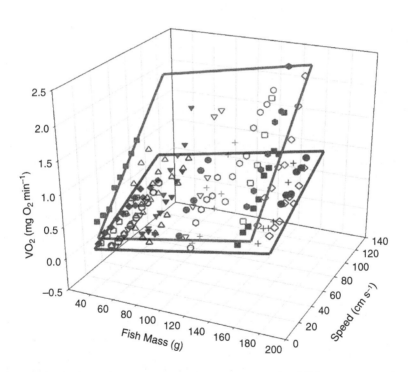

Fig. 4.7 A three-way plot showing the rate of oxygen consumption (VO_2) at two temperatures (more horizontal plane, 18°C; more inclined plane, 24°C) in relation to swimming speed and body mass for the chub mackerel, *Scomber japonicas*. Taken from Dickson et al. (2002) with permission from the Company of Biologists Ltd.

the aerobic muscle during acclimation to low temperatures appear to consti-
tute the driving force that enables carp to produce relatively high amounts
of mechanical power from their muscles and thus move quickly. This means
that the carp gets to have its cake and eat it too: if the need arises, it can move
quickly under adverse cold conditions, yet it does not pay a large energetic
cost because it can primarily use its aerobic muscles to power rapid locomotion. Acclimation is also a beneficial process for animals that live habitually in
very cold waters. The chub mackerel, *Scomber japonicus*, is a pelagic predator
that occurs in coastal regions throughout the world. In experiments in which
chub mackerel were allowed to acclimate to low (18°C) and high (24°C) temperatures, their rate of oxygen consumption was higher at the higher temperature, primarily because locomotion is more energetically expensive (Dickson
et al. 2002; see fig. 4.7). Chub mackerel also must recruit greater amounts of
slow-twitch aerobic muscle to drive locomotion at lower temperatures; thus,
for them, locomotion is more expensive at lower temperatures. This finding
provides some insight as to why these, and closely related fish such as tuna,
have physiological specializations to raise their body temperature. By raising
their internal body temperature several degrees higher than the temperature
of the ambient water, they can move aerobically for long distances at low energetic cost in cold waters.

In many cases, acclimation to one temperature may provide enhanced performance at that temperature (relative to an unacclimated state) but results in
diminished performance at other temperatures. For example, some fish species acclimated to low temperatures exhibit a boost in swimming performance at low temperatures but a decline in performance at higher temperatures.
Conversely, fish acclimated to high temperatures seem to enjoy a swimming
performance boost at high temperatures but undergo a decline in swimming
performance at low temperatures. By comparison, at least for the limited data
at hand, there appears to be little thermal dependence of running performance for endotherms. In ground squirrels, there is no significant decline in
either maximum sprint speed or the limb kinematics associated with running,
such as stride frequency, even as body temperatures range from 31°C to 41°C
(Wooden and Walsberg 2004). Why ectotherms and endotherms differ in how
temperature influences locomotion across a range of body temperatures is not
obvious.

There is also evidence that one cannot directly predict acclimation effects on
animal performance from acclimation effects on underlying physiology and
morphology. In some species there appear to be continuous changes in underlying physiology with acclimation temperature, whereas there appear to be
distinct thresholds in terms of locomotor performance in other species. This
difference indicates that temperature could have behavioral effects that cannot
be explained through effects on underlying morphology. There also appear to

be differences in the abilities of different developmental stages to effectively acclimate. Juveniles of some species possess the ability to modify their underlying phenotype in response to temperature, whereas this ability is diminished or lost in the adult stage. Finally, there are also significant seasonal influences on the ability of animals to acclimate. In general, animals can more easily tolerate cold temperatures at colder times of year, and hotter temperatures at hotter times of year, and thus vary seasonally in their ability to acclimate.

4.8 The ecological context of temperature

While it is important to consider the physiological effects of temperature on animal performance in the laboratory, it is the natural environment that has ultimately shaped the thermal specializations of animals. Ectotherms encounter a mosaic of temperatures in their natural environments, particularly for species that live in xeric regions. For example, in the Southwestern deserts of the United States, diurnal surface temperatures can reach as high as 50°C in the middle of the day. One of us (Irschick) remembers being at the Phoenix airport around midnight in 2005 and hearing on the radio that it was over 90°F (32.2°C)! While these high temperatures might make us crave air conditioning (particularly in my car, which lacked it!), the implications for animals are potentially more serious, as both extremely high and extremely low temperatures can impair body function or result in death.

This deleterious effect of temperature is manifested in a measure called the critical thermal maximum temperature, or CT_{max}. This value is the temperature at which ectotherms lose muscle function and, if the animal's internal body temperature achieves this temperature or higher for more than a few minutes, death can ensue. In general, species with high CT_{max} values can tolerate high temperatures, whereas those with lower values are more vulnerable to exposure to higher temperatures. This measure varies substantially among species and can be as high as 50°C for some species; but, for most ectotherms, it is usually around 40°C or less. Thus, most ectotherms, especially those in hotter environments, must either avoid hot days altogether or evolve anatomical, behavioral, or physiological mechanisms to maintain their preferred body temperature in the face of high ambient temperatures. As explained in Section 4.6, the majority of temperature-performance curves are asymmetric, with a strong decline at very high temperatures which are often near the CT_{max}. In short, overheating means poor performance, which could compromise an animal's ability to conduct vital ecological tasks. Operating at very low temperatures, by contrast, is usually less likely to be fatal over the short term but also is likely to impair performance.

However, it is also important to realize that some ectotherms have some control mechanisms for achieving desired body temperatures despite suboptimal ambient temperatures. These control mechanisms take various forms, but let's consider two examples. Some kinds of flying insects, especially large sphinx moths, which can be over a gram in mass, can achieve internal body temperatures that are much higher than those in the ambient environment (Heinrich 1971). Sphinx moths maintain, on average, thoracic temperatures of between 38°C and 42°C, regardless of the ambient temperatures that they are exposed to. For example, these moths can attain thoracic temperatures of around 40°C even when the ambient temperature is only 20°C, a doubling of temperature! Such high thoracic temperatures are necessary to achieve the high power outputs of the moths' flight muscles, thereby enabling them to rapidly move their wings at high frequencies of around 200 cycles per second (Hz). However, the ambient temperature is not always this high when the moths need to fly; thus, these animals need to have some internal mechanism to increase their internal body temperature. The moths solve this problem performing rapid contractions of their muscles before flight, also called physiological preflight "warm-up." This process has some similarities to shivering in humans; however, whereas the latter is involuntary, the former method is voluntary. In these moths, the flight muscles are contracted in simultaneous fashion (unlike during real flight when the muscles would alternate) until sufficient heat is generated; then, the pattern of muscular contractions changes to a pattern of alternate contractions, thus preparing the animal to fly. This behavior enables these moths and other flying insects to achieve high levels of flight performance, as sphinx moths are known for remarkable aerial acrobatics comparable to that performed by small birds! The cost of this behavior is substantial, however, and requires them to continually find food in the form of high-energy nectar to fuel their high metabolic rates. Variations of this method are employed by other flying insects as well, such as bumblebees.

A second common mechanism for achieving desired internal body temperatures in spite of suboptimal ambient temperatures is behavioral thermoregulation. A vast amount of work has examined this topic in desert lizards and has addressed how lizards achieve internal body temperatures that are desirable for rapid locomotor performance, as most lizards rely on rapid bursts of locomotion to elude predators and capture prey. All ectotherms can be placed roughly into two groups: thermoregulators or thermoconformers. Thermoregulators actively shuttle between cool and warm parts of the habitat and otherwise have an array of behaviors that allows them to maintain preferred body temperature, whereas thermoconformers are generally passive in relation to ambient temperatures. How often do ectotherms occupy thermal habitats that are clearly suboptimal for high levels of performance? This question has been investigated in detail in the thermoregulating desert lizard *Sceloporus merriami*. These lizards prefer exposed rocky hillsides in the Southwestern United States and northern Mexico that offer a range of thermal options. Bruce Grant and his colleagues have performed intensive thermal studies in which they studied how much time individual lizards spent in different thermal environments and then measured both the air temperature at those sites and the body temperatures of the lizards (Grant and Dunham 1988; Grant 1990). By extrapolating from studies relating temperature and sprint performance, they tested the percentage of the time that lizards spent experiencing "suboptimal" temperatures for sprinting and found that, on average, lizards spent about 90% of their time in thermal habitats in which they can sprint at or close to their maximum. Therefore, these lizards consciously chose to be in thermal habitats that allowed them to perform at high levels of locomotor performance, an ability that might prove handy in the instance of a predator attack or if a tasty insect comes crawling by. Does the precision of thermoregulation influence the ability to achieve high levels of locomotor performance in nature? Some animals are extremely precise thermoregulators, whereas other species are far more "sloppy," that is, have a broad set of preferred temperatures. Comparative studies with lacertid lizards show that those species that were "sloppy" thermoregulators were able to perform at high levels of maximum sprint speed only over a narrow range of body temperatures (Bauwens et al. 1995). In other words, sloppy thermoregulators are more likely to perform suboptimally, which might occur either because this kind of performance is not extremely important to them or because there is some other historical constraint or trade-off that causes them to perform poorly.

If temperature plays a central role in dictating animal performance, then it follows that, if one compares species that differ in their preferred body temperatures, there should be an evolutionary match between preferred body temperatures and the optimal temperature for performance traits. However, the available data are contradictory on this point. In some Australian skinks, there is not a match between the optimal temperature for sprint speed and preferred body temperature (Huey and Bennett 1987). Instead, although their optimal temperatures for maximum sprint speed are similar to the high values of their diurnal ancestors, these lizards have relatively low preferred body temperatures because of their nocturnal habits. Therefore, while preferred body temperatures appear evolutionarily labile, the optimal temperatures for locomotor performance are evolutionarily conserved. One explanation for this pattern is that the enzymes that control the muscles that drive performance are highly conserved and resistant to change; thus, nocturnal lizards may in many cases be at a locomotor disadvantage. However, a different study that examined *Anolis* lizards found the opposite result (Van Berkum 1986), namely, a positive evolutionary relationship between the preferred body temperature and their optimal temperature for sprinting. One possible explanation for these different findings is that the former study had a mix of nocturnal and diurnal lizard species, whereas the latter study included only diurnal lizards.

Further, the degree to which physiology and function are conserved in these two groups of lizards may differ as well.

References

Anderson CV, Deban SM. 2012. Thermal effects on motor control and in vitro muscle dynamics of the ballistic tongue apparatus in chameleons. Journal of Experimental Biology 215:4345–4357.

Andrews RD, Jones DR, Williams JD, Thorson PH, Oliver GW, Costa DP, Le Boeuf BJ. 1997. Heart rates of Northern elephant seals diving at sea and resting on the beach. Journal of Experimental Biology 200:2083–2095.

Angilletta MJ. 2009. Thermal Adaptation: A Theoretical and Empirical Synthesis. Oxford University Press, Oxford.

Bauwens D, Garland T Jr., Castilla AM, Van Damme R. 1995. Evolution of sprint speed in lacertid lizards: Morphological, physiological and behavioral covariation. Evolution 49:848–863.

Bennett AF. 1991. The evolution of activity capacity. Journal of Experimental Biology 160:1–23.

Bennett AF, Ruben JA. 1979. Endothermy and activity in vertebrates. Science 206:649–654.

Bergmann P, Irschick DJ. 2005. Effects of temperature on maximum clinging ability in a diurnal gecko: Evidence for a passive clinging mechanism? Journal of Experimental Zoology 303A:785–791.

Carrier DR. 1987. The evolution of locomotor stamina in tetrapods: Circumventing a mechanical constraint. Paleobiology 13:326–341.

Dawson TJ, Taylor CR. 1973. Energetic cost of locomotion in kangaroos. Nature 246:313–314.

Dickson KA, Donley JM, Sepulveda C, Bhoopat L. 2002. Effects of temperature on sustained swimming performance and swimming kinematics of the chub mackerel *Scomber japonicas*. Journal of Experimental Biology 205:969–980.

Gleeson TT, Hancock TV. 2001. Modeling the metabolic energetics of brief and intermittent locomotion in lizards and rodents. American Zoologist 41:211–218.

Gleeson TT, Hancock TV. 2002. Metabolic implications of a "run now, pay later" strategy in lizards: An analysis of post-exercise oxygen consumption. Comparative Biochemistry and Physiology part A: Molecular and Integrative Physiology 133:259–267.

Grant BW. 1990. Trade-offs in activity time and physiological performance for thermoregulating desert lizards, *Sceloporus merriami*. Ecology 71:2323–2333.

Grant BW, Dunham AE. 1988. Thermally imposed time constraints on the activity of the desert lizard *Sceloporus merriami*. Ecology 69:167–176.

Heglund NC, Cavagna GA, Taylor CR. 1982. Energetics and mechanics of terrestrial locomotion. III. Energy changes of the centre of mass as a function of speed and body size in birds and mammals. Journal of Experimental Biology 97:31–56.

Heinrich B. 1971. Temperature regulation of the sphinx moth, *Manduca sexta*. I. Flight energetics and body temperature during free and tethered flight. Journal of Experimental Biology 54:141–152.

Hoyt DF, Taylor CR. 1981. Gait and energetics of locomotion in horses. Nature 292:239–240.

Huey RB, Bennett AF. 1987. Phylogenetic studies of coadaptation: Preferred temperatures versus optimal performance temperatures of lizards. Evolution 41:1098–1115.

John-Alder H, Barnhart C, Bennett AF. 1989. Thermal sensitivity of swimming performance and muscle contraction in northern and southern populations of tree frogs (*Hyla crucifer*). Journal of Experimental Biology 142:357–372.

Johnson T, Bennett A. 1995. The thermal acclimation of burst escape performance in fish: An integrated study of molecular and cellular physiology and organismal performance. Journal of Experimental Biology 198:2165–2175.

Johnston IA, Temple GK. 2002. Thermal plasticity of skeletal muscle phenotype in ectothermic vertebrates and its significance for locomotory behavior. Journal of Experimental Biology 205:2305–2322.

Licht P. 1967. Thermal adaptation in the enzymes of lizards in relation to preferred body temperatures. In Prosser CL, ed. Molecular Methods of Temperature Adaptation. American Association for the Advancement of Science, Washington, DC, pp. 131–145.

Nielsen OB, de Paoli F, Overgaard K. 2001. Protective effects of lactic acid on force production in rat skeletal muscle. Journal of Physiology 536:161–166.

Owerkowicz T, Farmer CG, Hicks JW, Brainerd EL. 1999. Contribution of gular pumping to lung ventilation in monitor lizards. Science 284:1661–1663.

Robin ED, Lewiston N, Newman A, Simon LM, Theodore J. 1979. Bioenergetic pattern of turtle brain and resistance to profound loss of mitochondrial ATP generation. Proceedings of the National Academy of Sciences 76:2922–3926.

Rome L, Funke RP, Alexander R McN. 1990. The influence of temperature on muscle velocity and sustained performance in swimming carp. Journal of Experimental Biology 154:163–178.

Schmidt-Nielsen K. 1972. Locomotion: energy cost of swimming, flying, and running. Science 177:222–228.

Taylor CR, Heglund NC, Maloiy GM. 1982. Energetics and mechanics of terrestrial locomotion. I. Metabolic energy consumption as a function of speed and body size in birds and mammals. Journal of Experimental Biology 97:1–21.

Tobalske BW, Hedrick TL, Dial KP, Biewener AA. 2003. Comparative power curves in bird flight. Nature 421:363–366.

Tucker VC. 1970. Energetic cost of locomotion in animals. Comparative Biochemistry and Physiology 34:841–846.

Van Berkum FH. 1986. Evolutionary patterns of the thermal sensitivity of sprint speed in *Anolis* lizards. Evolution 40:594–604.

Weibel ER, Bacigalupe LD, Schmitt B, Hoppeler H. 2004. Allometric scaling of maximal metabolic rate in mammals: Muscle aerobic capacity as determinant factor. Respiratory Physiology and Neurobiology 140:115–132.

Weinstein RB, Full RJ. 1999. Intermittent locomotion increases endurance in a gecko. Physiological and Biochemical Zoology 72:732–739.

Williams TW. 1999. The evolution of cost efficient swimming in marine mammals: Limits to energetic optimization. Philosophical Transactions of the Royal Society of London B 1380:193–201.

Williams TM. 2001. Intermittent swimming by mammals: A strategy for increasing energetic efficiency during diving. American Zoologist 41:166–176.

Williams TM, Haun JE, Friedl WA. 1999. The diving physiology of bottlenose dolphins (*Tursiops truncates*). I. Balancing the demands of exercise for energy conservation at depth. Journal of Experimental Biology 202:2739–2748.

Wooden MK, Walsberg GE. 2004. Body temperature and locomotor capacity in a heterothermic rodent. Journal of Experimental Biology 207:41–46.

Zani PA, Gottschall JS, Kram R. 2005. Giant Galápagos tortoises walk without inverted pendulum mechanical-energy exchange. Journal of Experimental Biology 208:1489–1494.

5 | The evolution of performance I
Mechanism and anatomy

5.1 Mechanistic limits on performance

Whereas performance capacities emerge from the whole organism, their expression is largely based on the underlying structural and physiological aspects of the phenotype. In other words, function typically has a direct connection with form and, while not always observed, it serves as a guiding principle. A cogent example is the relationship between muscle physiology and aerobic capacity, more commonly known as endurance. Within vertebrates, there are several muscle types, but two important ones are slow-twitch and fast-twitch fibers. Slow-twitch fibers, also known as "red muscle," or the "dark" meat that one fights over at Thanksgiving, contain large numbers of mitochondria for efficient oxygen transport and can perform low-force contractions for relatively long periods of time before fatigue sets in. You might use slow-twitch muscles if you take a walk around your neighborhood. By comparison, fast-twitch fibers, or "white muscle," produce more force but only for relatively short periods. The next time you take out a heavy bag of trash or run after your dog, you will be likely be using your fast-twitch muscles in large supply. Animals with high aerobic capacity tend to have relatively large amounts of red muscle, whereas animals that rely on short, explosive bursts of movement tend to have relatively more fast-twitch muscle. The relative amount and distribution of these muscle types represent the building blocks for most kinds of animal performance.

Many factors are important for determining how well and for how long animals can perform various tasks. Some examples include the position and sizes of individual muscles, and the shapes and sizes of bones, tendons, ligaments, and nerves, among others. Because there is variation among and within species in anatomical structure and physiology, taking an evolutionary approach will allow us to address the drivers of animal performance at a broad scale (Lauder 1990; Alexander 2003). By doing so, we can ask whether animals can evolve anatomical structures or other features that allow them to exhibit high performance capacity in suboptimal conditions, such as when ectotherms live in low temperatures. A very recent example is the opah, *Lampris guttatus*, a marine fish found off the coast of Southern California. This species exhibits whole-body endothermy, a first among fishes (Wegner et al. 2015). It does this with a combination of countercurrent heat exchangers within its gills and constant "flapping" of its wing-like pectoral fins. These warmer temperatures occur throughout the body, including the heart, and ultimately enhance physiological function and the ability to capture prey in cold water.

A mechanistic and evolutionary approach will also allow us to understand the nature of trade-offs. Can animals evolve morphological features that enable them to perform well in many different environments, or are there irreconcilable trade-offs that are imposed by the basic nature of animal form? Finally, has the evolution of anatomical structure allowed animals to overcome constraints imposed by their ecology or their ancestry, thereby acting as an innovative and compensating force? However, with the exception of chapter 11, this book will not examine the role of exercise for modifying physiology or anatomy, as this effect seems somewhat confined (as far as we know) to humans and a subgroup of animals. Further, whereas exercise or training can improve some aspects of organismal performance, such as endurance, the basic structure and physiology of bones, muscles, and tissues is typically molded over millions of years to match the functional needs of each species for survival in their habitats (Lauder 1990, 1996; Alexander 2003). In at least some cases, such characteristics appear resistant to exercise training. In general, the amount of variation in anatomical structure, and its consequent effects on animal performance, varies far more among animal species than within a single species or individual, regardless of the amount of physical training that is undertaken.

5.2 The mechanism and anatomy of trade-offs

One of the best ways to understand the mechanistic underpinnings of animal performance is to consider the pervasive role of trade-offs, a topic that is explored in more detail in chapter 7. In a functional sense, trade-offs occur when there is a conflict between the simultaneous optimization of two or more performance traits (Huey and Hertz 1984). These trade-offs are ubiquitous and define much of the evolutionary possibilities available to organisms. Trade-offs occur for several reason, including resource limitation, limited physical space, or mechanical trade-offs imposed by basic physical laws. Trade-offs are especially relevant during development, when organisms begin to assemble the morphological patterns and parts that will ultimately form the phenotype. It is during this developmental phase when energetic investment is typically most limited. Second, the mechanical demands of different physical structures often conflict with one another. This results in some phenotypes being well suited for high performance in some conditions but not others. An oft-used example is the use of slow-twitch muscles that are well suited for aerobic activities, and the use of fast-twitch muscles for movements that are rapid, intense, or both.

Trade-offs among performance traits are ubiquitous at the whole-organism level, and they are often driven by conflicting demands of anatomical structures. Given that we will discuss trade-offs in more detail in chapter 7, in this chapter we will present only one example in the context of the evolution of anatomical trade-offs and feeding. Feeding is a good system in which to study these trade-offs because there are often conflicting demands for biting rapidly, such as when capturing elusive prey, and biting forcefully, such as when crushing hard prey. The specializations that animals possess for consuming ants show some trade-offs when compared with feeding on more typical prey. The occasional consumption of ants is widespread, but some specialized mammals and lizards, among other animals, subsist almost entirely on a diet of ants, such as the mammalian anteater, suborder Vermilingua, and various species of horned lizards, genus *Phrynosoma*, that occur in the southwestern deserts of North America. At a first glance, a diet consisting nearly entirely of ants poses problems. Although ants are often numerous, they are not highly nutritious compared to other insects and are sometimes toxic; thus, specialized ant-eating animals must consume large numbers of ants but yet avoid being poisoned. To do so requires significant evolutionary modifications of the feeding structures present in closely related species that do not specialize on ants.

The jaws of most mammals and lizards are designed for some form of crushing or mastication, but neither process is useful for an exclusively ant-based diet, as most ants require little processing, but large numbers must be consumed relatively quickly. Both anteaters and horned lizards appear to have resolved this conundrum in a similar manner, by a reduction of the overall complexity of the feeding structures (Naples 1999; Meyers and Herrel 2005). When a mammal or a lizard feeds, two variables are especially important for processing the prey: gape duration, that is, the duration of the opening and closing of the jaws, and bite force. How hard an animal can bite depends greatly on the size of the object it is feeding on, as feeding on large prey items greatly reduces mechanical advantage and thus diminishes bite force. If a lizard or a mammal were to masticate a prey item that was large, hard, or both (i.e., chew it up into little bits to facilitate swallowing), both long gape durations and high bite forces would be required. On the other hand, when consuming large numbers of very small prey, bite force is unimportant, except possibly for very large and hard ants. Further, gape durations should ideally be short when consuming ants, as little or no mastication is required. In short, there should be a trade-off for the kinematic and performance traits for the jaws of specialized ant-eating species when compared to species which eat a greater diversity of prey types. Unfortunately, the mammalian anteaters are so unique that it is difficult to compare them with other species, but horned lizards are a good model system because, while some species exclusively eat ants, other species consume a wider variety of prey, although ants remain an important diet item for these species. As an aside, it's worthwhile mentioning that diets in animals rarely remain static, and there is evidence that horned lizards enjoy a more diverse insect diet in the spring, when more insects are available, and narrow their diet during the hot and lean summer months (J. Meyers, personal observation).

Among horned lizard species, there have been dramatic reductions in key muscles used to masticate prey, resulting in ant specialists that possess a more gracile head well suited for rapidly seizing prey (Meyers and Herrel 2005; Meyers et al. 2006). Interestingly, while the muscles controlling the closing of the jaw have been reduced, those involved in jaw opening remain prominent in ant specialists. The reduction in closing muscles in turn results in ant specialists having relatively low bite forces, whereas horned lizards species with a more catholic diet bite much harder. However, the relative reduction and enlargement of muscles involved in jaw closing and opening is only part of this story. Within lizards, ant specialists have evolved several times, and this independent origin has resulted in different strategies for capturing sufficient numbers of ants to survive. The Australian moloch lizard, *Moloch horridus*, belongs to the lizard family Agamidae (unlike horned lizards, which belong to the family Phrynosomatidae) and is a superb ant specialist. The most noticeable of its specializations for ant eating is its spiny body, which renders it largely invulnerable to predators that would feed on normally feed on sedentary lizards. When one compares the moloch with a typical insect-eating lizard (the bearded dragon, *Pogona vitticeps*), several differences in the kinematic mechanisms of feeding emerge (Meyers and Herrel 2005). First, the moloch is distinctive in exhibiting a remarkably fast gape duration, which is aided by rapid projections of their tongue to quickly pluck ants from the ground, whereas the bearded dragon more ponderously grabs, chews, and swallows prey. The rapid and almost machine-like use of the tongue is unique even when compared to distantly related ant-eating horned lizards, which consume ants by quickly grabbing and biting ants that appear before them.

Second, based on these values of gape durations, it is possible to calculate how long it would take for each of these species to consume enough ants to survive. Because bearded dragons have much longer gape durations and tend to masticate their prey, they would take an extraordinary 8 h or more to consume enough ants to survive! By comparison, both horned lizards and the moloch would take only about one-fourth the time to consume the equivalent number of ants, thereby leaving the remainder of the day to engage in other lizardly activities, such as basking, mating, or patrolling their habitat. In all, lizard ant specialists appear to have sacrificed bite force to enable faster bites, thus facilitating the consumption of large numbers of ants. This evolutionary modification occurred by a reduction in key feeding muscles in the head, thereby enabling the head to be more gracile. These traits also appear to have independently evolved, strongly suggesting that they are adaptive.

These principles can also be observed across a variety of other organisms. As noted above, the roles of red and white muscle in different kinds of performance has been well studied. Species that rely on extended bouts of movement often possess relatively large amounts of red muscle that fuel their impressive aerobic capacities (Greer-Walker and Pull 1975; Rome et al. 1988). For species that rely on quick bursts of speed to capture prey, there is a greater specialization toward anaerobic locomotion fueled by relatively large amounts of white muscle. For example, some pelagic fish species, such as tuna, and some sharks will migrate long distances at relatively fast speeds and therefore possess large amounts of red muscle, whereas more sedentary fish species tend to possess less red muscle and more white muscle (Greer-Walker and Pull 1975). However, many animals often require capacities for high endurance and rapid movements and have a complex mosaic of red and white muscles that can be sequentially recruited to fulfill these ecological needs.

Given the different requirements of white and red muscle, and their different roles in aerobic and anaerobic locomotor performance, one might expect a trade-off between the ability to run at high speeds and the ability to run with high endurance. Among lacertid lizard species, there is a trade-off between maximum speed and endurance, such that fast lizard species have poor endurance, and vice versa (Vanhooydonck et al. 2001). However, when different individuals are compared, either within humans or within garter snakes, there is little evidence for a trade-off between maximum speed and endurance (Garland 1988, 1994), with some exceptions, such as in cod, *Gadus morhua* (Reidy et al. 2000). One explanation for these conflicting results is that anatomical differences among individuals are too modest compared to the myriad of other factors such as motivation or other physiological variables that affect performance traits, such as maximum speed and endurance. Among divergent species that differ in the relative proportions of muscle fibers and even their anatomical structure, variation in performance due to other factors may be inconsequential compared to these overarching factors. More studies linking the exact muscle physiology and performance across individuals and species might resolve these conflicting results.

From a mechanistic perspective, anatomical structure clearly plays a role for creating trade-offs among performance traits in many animal species, but there is much that we don't understand about why anatomically imposed trade-offs are manifested in some systems but not others. With this in mind, let's explore some other examples for how anatomical structure acts as a key foundation of performance in animals.

5.3 Variation in anatomical structure as a foundation of animal performance

As explained in chapter 1, anatomical structure refers to the three-dimensional configuration of morphological parts, such as bones, ligaments, tendons, muscles, and other tissues. The manner by which such parts interact with one another allows animals to produce mechanical power for dynamic actions. For instance, simple motions of the hindlimb in tetrapods, such as during locomotion, are driven by several large muscles, such as the quadriceps, and the neurons that innervate those muscles. When a cat jumps, for example, the quadriceps relaxes and then explosively contracts just prior to jumping; however, other anatomical structures also play a role, including the connective tissues of the knee and foot, as these tissues can compress like a spring to provide power for the jump. In other words, to understand why some animals can jump farther or run faster than others, we need to understand the anatomical structures that drive the movement or activity in question. Although one could write a dozen books (people have!) on how nearly every feature of organismal structure and physiology influences performance traits, here we will focus on a few salient examples, primarily involving larger aspects of anatomical structure (muscles, tendons, and bones) that explain certain basic principles.

For vertebrates, although the actual dynamics of force production are complex, the primary structures that provide energy for dynamic movement are muscles and elastic tissues such as tendons (fig. 5.1). Muscles contract and produce mechanical power, thereby moving objects, whereas elastic tissues can store elastic energy like springs in some animals and therefore convert potential energy to kinetic energy. Bones and soft tissues other than tendons act as the foundation for movement, as tendons typically attach muscle to bones. For many years, the prevailing view was that muscles were the primary agents driving movement and, in many cases, this view still holds. Muscles provide power, but at a cost, as each cycle of muscle contraction consumes energy in the form of ATP. However, there is increasing evidence that elastic elements can store and release energy much like a spring, especially for larger

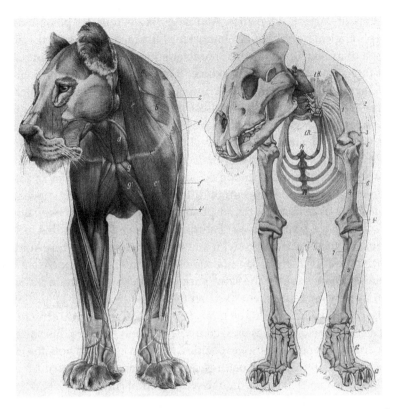

Fig. 5.1 A drawing showing the configuration of muscles, tendons, soft tissues, and bones for a large vertebrate (lion). Image from Wikimedia Commons.

animals such as deer or kangaroos, which can take advantage of gravity to compress large and springy tendons (Blickhan 1989; Farley et al. 1993). Unlike muscles, elastic elements impose no energetic cost but require an input of initial energy in order to be stretched or "loaded." This initial energy can be freely obtained using the force of gravity, such as when a kangaroo jumps and compresses tendons in its limbs. Bones also play an important role, as they are the anchors for the powering muscles and tendons, although they do not provide energy per se for movement. While one can potentially dissect the roles of each of these elements, it is wise to remember that anatomical structures work in tandem with one another in a closely coordinated fashion, and it is this integration of different parts that is essential for driving different kinds of performance capacity.

However, it is also important to examine aspects of structure within each of these elements. As noted above, muscles exist in several varieties, including the fast- and slow-twitch types already mentioned, as well as other kinds that are intermediate between these extremes, and contain elements designed for movements of varying intensities. At a larger scale, muscles possess distinct-ive morphological types that are important for function, such as the angle of pennation of muscle fibers, as this characteristic can influence the amount and kind of force that muscles can produce. Similarly, bones vary in their struc-tural elements in ways that directly impact performance capacity. Bones are generally hollow structures that possess trabeculae which act like buttresses, thereby strengthening the bone but still preserving their lightweight nature (would you want to move with heavy bones?). However, because bones are designed for withstanding loads, their cross-sectional thickness and ability to resist stress forces varies among ecologically different species of the same size, and even more so for species of different sizes. For example, we usually think of cartilage as being soft and pliable (just wiggle your nose); however, in sharks and rays, which have substituted cartilage for bone, this tissue has been stiffened to allow them to swim and feed. In stingrays that consume hard prey such as clams, the cartilaginous jaws are hardened by extensive calcifi-cation and possess some of the same supportive features as bone (trabeculae, fig. 5.2), thereby enabling them to crush their prey (Summers et al. 1998). We could list other examples, but the main point is that anatomical structure var-ies in scale, dimension, and overall composition, and much of this variation is tied to variation in performance capacities. Below we discuss how the creative arrangement of anatomical structures can enable animals to perform at high levels of performance in suboptimal environmental conditions.

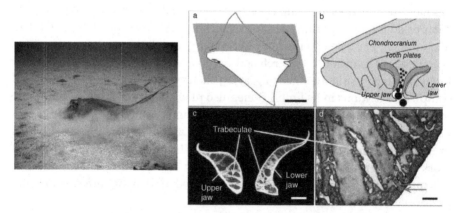

Fig. 5.2 *Left*, An image of a stingray. *Right* (*a–d*), Figure showing various views of the trabecu-lae from stingray jaws. Stingrays are elasmobranchs and are therefore cartilaginous, but they also consume hard prey, such as clams. The trabeculae assist in allowing stingrays to consume such hard prey. Image on right from Summers et al. (1998). Stingray image is from Wikimedia Commons.

5.4 Creative arrangement of anatomical structure to enable high performance in suboptimal settings

One of the best ways to understand the role of anatomical structure for animal performance is to examine how some species can produce high performance in clearly disadvantageous settings. For example, muscles generally perform better when they are at their optimal temperature, and suboptimal temperatures are problematic for ectotherms that reside in cold waters. Nonetheless, some animals typically considered as ectotherms must exert high levels of locomotor performance to effectively capture prey. The mackerel sharks, order Lamniformes, which includes the well-known great white shark, *Carcharodon carcharias* (fig. 5.3) and the lesser known mako shark, genus *Isurus* (fig. 5.3) each hunt elusive prey, such as marine mammals for great white sharks, and bluefish and other pelagic fish for mako sharks, that require the sharks to generate rapid bursts of speed. While great white sharks are the fast and powerful Ford F-350s of the ocean world, mako sharks are more like sleek Porsches, capable of moving at extremely fast speeds. The hunting strategy of the white shark has been well studied and highlights the functional challenge of moving in their aquatic habitat. In at least some parts of their range, adult great white sharks hunt by cruising below the water surface while looking for silhouettes above. Once prey is spotted, the sharks rapidly propel themselves upward and aim to severely injure or kill the prey by a single traumatic bite (Klimley et al. 2001) Therefore, the functional challenge for large great white sharks (they can exceed 2,000 lbs.) is to propel themselves at very high speeds directly upward in cold water. The movement of such a large mass requires particularly high muscle power outputs, which would normally be difficult for a muscle operating at a suboptimal temperature to produce. In addition to

Fig. 5.3 *A*, Image of a mako shark, *Isurus oxyrinchus*. *B*, Image of a great white shark, *Carcharodon carcharias*. Both of these shark species employ a sophisticated countercurrent system, as well as other thermal specializations, to ensure that their "core" is substantially hotter than the cold water they frequent. The mako shark image is from Bill Fisher—333 Productions. The white shark image is from Neil Hammerschlag.

the need for high bursts of speed, great white sharks also are known to travel long distances, often into deep and cold ocean waters far from the warmer coastal waters in which they are most frequently encountered (Bonfil et al. 2005). Thus, great white sharks must be able to perform both short bursts of intense anaerobic locomotion to capture prey, and longer bouts of sustained aerobic locomotion to travel across wide swaths of the ocean, all while operating in cold waters that can vary in temperature from 12°C to 24°C. These ambient temperatures are usually less than the optimal temperatures for rapid limb movements, given that such temperatures are generally above 25°C. As the muscles of these sharks are not especially efficient at cold temperatures, the sharks must therefore somehow raise the internal temperature of their muscles by a large margin.

Both kinds of sharks possess several anatomical specializations that allow them to exhibit high levels of locomotor performance in suboptimal conditions. First, each possesses a countercurrent thermoregulatory system that consists of a net of arteries, capillaries, and other blood vessels (the rete mirabile, or "marvelous web") and which allows heat to be effectively removed from the cold oxygenated blood flowing from the gills by the hot, oxygen-poor blood flowing from the heart. The rete mirabile appears to be especially prominent near the head (for heating the brain), the viscera (for enhancing digestion, which is especially important for digesting fatty prey, such as sea lions), and elongated muscles that run alongside the trunk and act as the primary propellers of movement. This countercurrent system, along with several other

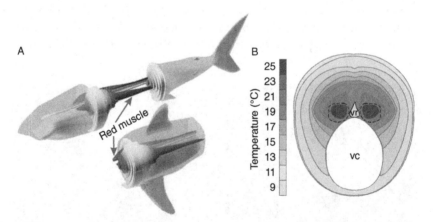

Fig. 5.4 Figure showing the high internal "core" temperatures of a laminid shark, the salmon shark, *Lamna ditropis*, which is a close relative of the great white and the mako. *A*, Diagram showing the location of red muscle within the shark. *B*, The temperature distribution within the shark. Note that the internal core temperatures can reach about 25°C. This temperature allows the red muscle that is a key driver of movement in these high-performance sharks to stay "hot" in cold waters. Taken from Bernal et al. (2005) with permission from Nature Publishing group.

features, such as their large mass and relatively high metabolic rate, allows great white sharks and mako sharks to attain core body temperatures about 10°C –13°C degrees higher than the external water (Bernal et al. 2005), resulting in "regional" endothermy, in which the viscera, brain, and primary locomotor muscles are "hot" while the periphery is "cool" (fig. 5.4). Unlike sharks and as noted in section 5.1, a recent study found that opah, *Lampris guttatus*, a marine fish found off the coast of Southern California, has whole-body endothermy (Wegner et al. 2015). This situation is more similar to that in mammals, which have a fairly constant temperature throughout the body.

The increased temperature in sharks works in conjunction with the muscular anatomy. Both kinds of sharks propel themselves with rapid and powerful side-to-side strokes of their caudal fin, which is powered by these elongated muscles that span much of the length of the trunk and the muscular tail. The muscle anatomy of mako sharks has been well studied, and there is also evidence that the structure and function of their elongated muscles is convergent to that in tuna, one of their primary prey, even though these species are distantly related (Donley et al. 2004). These muscles allow both mako sharks and tuna to swim quickly by acting as pistons to propel the tail back and forth rapidly. The efficiency of these muscles is enhanced by the high core temperatures generated by the countercurrent temperature exchange system. An increase of about 10°F provides about a threefold boost in overall speed. However, this lifestyle also imposes costs in the form of relatively high food requirements, which is one of the reasons why great white sharks focus on energy-rich prey such as sea lions, which possess large amounts of fatty blubber, and mako sharks consume bluefish or fish that are similarly energy rich. This shark example shows how quite sophisticated and creative arrangements of anatomical structures can evolve in the presence of evolutionary pressures to perform at high levels in suboptimal environments. Despite this example, it is important to remember that, for the majority of smaller ectothermic animals, it is more difficult to overcome the pervasive effects of temperature and that most that reside at cool temperatures possess relatively slow metabolic rates and are relatively sedentary.

Just as high locomotor performance at low temperatures poses significant problems, performing any kind of locomotion in the absence or near-absence of oxygen poses perhaps an even more daunting challenge, especially when extended movements are used. Aerobic metabolism requires the input of oxygen and, while anaerobic metabolism can be sustained for a short period of time, the buildup of deleterious waste products such as lactic acid can constrain the amount of time for which this mode can be used, at least for most animals. This dynamic poses a problem for air-breathing diving animals, particularly marine mammals. Marine mammals are known to dive for extended periods of time, in many cases exceeding the predicted "dive limit" based on calculations of lung volume, metabolic rate, and so forth. While scientists don't completely understand the reasons for why some marine mammals can exceed these theoretical limits, investigations of the anatomy show several specializations for spending relatively long periods of time underwater in a hypoxic state (Kanatous et al. 2002). In the case of some marine mammals, such as northern fur seals, *Callorhinus ursinus*, and Steller sea lions, *Eumetopias jubatus*, their muscles show an increased numbers of mitochondria and increased citrate synthase activity and a greater capacity for oxygen diffusion and storage (due to an increased myoglobin concentration), as well as other changes. Altogether, these modifications enable these species to increase the total amount of time they can spend underwater by increasing their ability to extract oxygen from their initial oxygen stores.

Among marine mammals, the Weddel seal, *Leptonychotes weddellii*, is a particularly outstanding diver, performing dives exceeding 60 min. By comparison, the northern fur seals, *Callorhinus ursinus*, and Stellar sea lions, *Eumetopias jubatus*, typically dive for only several minutes. When one compares the muscle physiology of Weddell seals with these other seal species, the Weddell seals seem to exhibit some but not all of the various muscular specializations. Indeed, the volume of mitochondria, which are used for oxygen metabolism, for the long-diving Weddell seals was comparable to that in a typical terrestrial mammal, such as a dog (fig. 5.5). Part of the answer to this riddle may be that the Weddell seal musculoskeletal system is composed almost exclusively of slow-twitch muscle fibers that are specialized for low-intensity aerobic movements; therefore, this seal may be able to dive for long periods of time by moving at low intensities, thus decreasing the overall demand for oxygen. Further, the use of intermittent locomotion, in this case, gliding during ascent, may minimize oxygen requirements (Williams 2001).

Much of the aerobic performance advantage of marine mammals is derived from the dive response, a physiological syndrome that is triggered during deep dives (Davis et al. 2004). During the forced submergence of a marine mammal, blood oxygen is sequestered to the brain and heart, thus causing peripheral tissues to become anaerobic. This response protects the animal from dying of asphyxiation. However, therein lies the rub: while this response does a marvelous job of keeping marine mammals alive, it also seems to run counter to the need for rapid locomotion. In other words, marine mammals must be active at the exact same time that their metabolism shuts down. In free-diving marine mammals, this response still exists but appears to be less intense than for animals undergoing forced submergence; thus, free-diving marine mammals are able to maintain aerobic metabolism in their peripheral tissues despite their oxygen-starved state. Therefore, the primary function of the dive response, during which blood flow to peripheral tissues is greatly reduced, the rate of heartbeats declines, and breathing largely ceases, may be to regulate

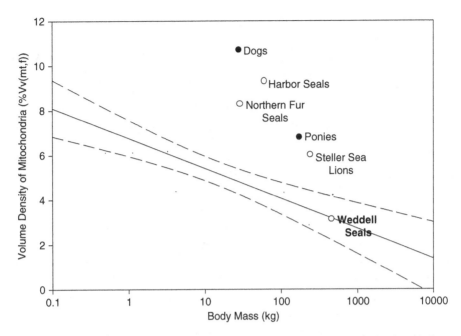

Fig. 5.5 A plot showing the relationship between body mass and the density of mitochondria for various animal species, including several seal species. Taken from Kanatous et al. (2002) with permission from the Company of Biologists Ltd.

how much hypoxia occurs in the skeletal muscles that drive locomotion. In addition, marine mammals appear to possess morphological and enzymatic adaptations that improve the intracellular diffusion of oxygen when muscle tissues are exposed to low levels of oxygen. The dive response and the corresponding morphological and physiological specializations of marine mammals enable them to dive for long periods in the presence of little oxygen.

5.5 Anatomical structure as a way of overcoming physical limits

So far, we have spent some time discussing ways that animals overcome environmental constraints on performance, such as low temperature or low levels of oxygen. Another potential constraint that must be overcome consists of intrinsic limits that are either self-imposed, such as when limitations may be necessary for survival in a particular habitat, or due to historical accident. One limitation involves the ability to grab and manipulate prey. For animals such as frogs that typically rely almost exclusively on their tongues to project and capture prey, the functional challenge of propelling the tongue with both accuracy and speed is daunting and has resulted in some remarkable specializations.

There are also extremes along which animals have evolved. On the one extreme, slow-moving toads rely on rapid and accurate projections of the tongue to capture prey. On the other hand, frogs that can jump long distances rely on a combination of locomotion and tongue projection to capture prey. One of the most interesting examples of how mechanistic changes in anatomy and behavior have influenced performance comes from studies of feeding in frogs and toads (Nishikawa 2000). Within this group, three primary ways of feeding have evolved: mechanical pulling, inertial elongation, and hydrostatic elongation. In mechanical pulling, the frog leaps forward, grabs the prey with its mouth, and shortens the tongue. In this mode, the velocity and acceleration of the tongue is relatively modest, and accuracy is high (Table 5.1). In inertial elongation (used mostly by toads), the toad projects the tongue at high velocities and accelerations (about three to four times higher than for mechanical pulling), yet the strike itself is not particularly accurate, hitting the prey only about a third of the time (Nishikawa 1999). Finally, hydrostatic elongation similarly consists of elongation of the tongue but does so by hydrostatic means in which the tongue elongates to about 200% of its resting length at relatively slow velocities and accelerations (about 10% relative to inertial elongation) but with high accuracy (>99%). Broadly, these three modes seem to represent a trade-off between accuracy and the speed of tongue movement, with those toads that are most accurate having the lowest velocities and accelerations, and vice versa. These trade-offs are closely linked to the lifestyles of the species that have evolved them. Toads are generally slow and usually rely

Table 5.1. A summary of various performance parameters for different kinds of feeding modes in frogs. Taken from Nishikawa (2000) with permission from Elsevier Press.

Characteristics	Mechanical Pulling	Inertial Elongation	Hydrostatic Elongation
tongue movement	shortens	elongates to 180%	elongates to 200%
velocity (cm/s)	15–50	250–400	~24–240
acceleration (m/s^2)	1.5–9.5	>310	~3.5–145
tongue/jaw synchrony	no	yes	no
aiming	distance	head only	distance, azimuth, elevation
accuracy	~95%	~33%	>99%
on-line correction	yes	no	yes
feedforward control	yes	yes	yes
feedback control	yes	no	yes
hypoglossal afferents	no	yes	no

Fig. 5.6 A phylogenetic tree depicting the evolution of the mechanisms for tongue-projection in frogs. Taken from Nishikawa (1999) with permission from Royal Society Publishing.

on toxic skin secretions to avoid predators, and rapid tongue projection allows a slow animal to capture quick prey. By contrast, frogs that use mechanical pulling rely more on rapid locomotion driven by their long hindlimbs, which are well suited for jumping, to capture prey. Each of these three modes of feeding have arisen independently multiple times within frogs and toads (fig. 5.6), providing an outstanding opportunity to determine whether the evolution of new feeding modes occurs in concert with evolution of anatomical changes.

These different feeding performance strategies have evolved alongside substantial modifications in the sizes and shapes of key muscles, connective tissues, and other structures. One transition occurred between mechanical pulling and inertial elongation and coincided with several morphological changes, including a less massive tongue, increases in the lengths and contraction velocities of some muscle fibers, alterations in the insertion points of key muscles, and a decline in the amount of connective tissue. Connective tissue appears to diminish tongue elongation in mechanical pullers. Other evolutionary transitions, such as that from inertial to hydrostatic elongation, resulted in fewer morphological modifications: the latter change resulted in only the addition of some new muscular compartments, and even less connective tissue in the tongue.

Not only are there differences in anatomical structure among these different kinds of functional groups, but the motor control of their muscles appears to differ as well. First, in species in which the tongue must be highly accurate, such as in inertial elongators, there must be close coordination between the fast jaw movements and tongue. By comparison, this kind of coordination is not needed in other feeding modes, such as in mechanical pullers. Second, the different species use different kinds of feedback between muscles and nerves. For example, mechanical pullers and hydrostatic elongators depend on feedforward and feedback control of tongue movements, whereas inertial elongators only employ feedforward, open-loop control to control coordination between movements of the jaw and tongue.

In other sedentary animals that use rapid tongue projections to capture prey, there are also interesting morphological specializations that enhance feeding performance. Studies of the motor control of feeding in plethodontid salamanders have also examined the question of whether among-species variation in performance capacity results in alternations in motor control strategies for the driving muscles. Salamanders of the family Plethodontidae, a diverse and speciose group, all project their tongues to some extent to capture prey, but there is variation in the degree of elongation and in how muscles control this behavior. In two groups, the Desmognathinae and the Plethodontinae, the salamanders have protrusible tongues and attachments between the tongue and the lower jaw to constrain tongue projection. Two other groups, the Bolitoglossini and most species within the Hemidactyliini, appear to have undergone an evolutionary modification that frees up the constraint on tongue projection; they have lost the attachments between the tongue and the lower jaw and can extend their tongues much farther—in some species, to nearly 80% of the animal's body length—and at high velocities and accuracies (Deban and Dicke 1999). There are three primary muscles that are responsible for this mode of feeding: the tongue retractor, the tongue protractor, and the jaw depressor, each of which play a distinct role during feeding. In salamander species from each of these above groups, those with short, attached tongues showed a pattern of simultaneous activation. By comparison, for those species with longer, freer tongues, the typical pattern was nonsimultaneous activation. It appears that the motor pattern for highly nonsimultaneous activation of antagonistic tongue muscles may have evolved alongside free, elongated tongues in these salamanders. This means that the motor control of salamanders is far more variable and less constrained than is typically envisioned. In other words, to meet the need for different kinds of feeding performance, there has been significant evolution in both muscle structure and muscle architecture, as well as motor function.

Another physical limitation is gravity, and the ability of several animal groups to fly is remarkable given the inherent challenges (fig. 5.7). Flight can be energetically expensive and requires a substantial amount of muscular power because the animal is working against gravity. In terrestrial animals, the relationship between metrics such as speed or total distance traveled per unit time is usually related linearly to mechanical power. If you want to run faster or farther, you have to produce more power. However, for flight, the relationship between flight speed and power is more complicated because of the complex dynamics of wind flow over the wings and its relation to flapping flight (Dial et al. 1997). Indeed, the relationship between power and speed varies among studies and among species. Some factors include the shape of the bird wing, and flight style. Those species, such as magpies, that can modify the way that they use their wings, for example, by moving them in an intermittent fashion, may not have to produce more power at higher speeds. Other species that are less able to modify how they use their wings during powered flight must produce more power as they fly faster. One of the factors that

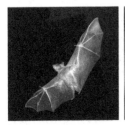  

Fig. 5.7 Bats (*left*), insects (*middle*), and birds (*right*) have each independently evolved the ability for powered flight using very different pathways. Images from Wikimedia Commons.

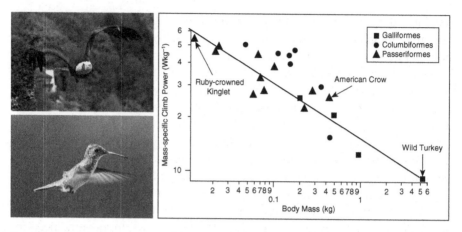

Fig. 5.8 *Top left*, Image of an eagle flying. *Bottom left*, Image of a hummingbird in flight. *Right*, Plot showing the negative relationship between body mass and the ratio of power to mass, for several bird species. Small birds, such as hummingbirds, possess far more power per unit of mass than large birds, such as eagles, do. Plot taken from Dial et al. (2008) with permission from Cell Press. The hummingbird and eagle images are from Wikimedia Commons.

influence how much power birds can produce is body size. In general, mass-specific power output declines as birds become larger; that is, small birds can fly at higher accelerations and are more maneuverable than large birds (Dial et al. 2008), an observation that seems obvious if you compare a humming-bird and an eagle (fig. 5.8). Why power output declines is complex, but one factor in relation to flying is the manner by which wingbeat frequency scales with body size, as smaller birds have higher wingbeat frequencies than larger birds. When the amount of muscular power generated during takeoff flight is estimated for bird species of varying sizes, it scales slightly negatively with the mass of the pectoralis, which is one of the large muscles used to propel the bird wing; however, the scaling coefficient is less than that for wingbeat frequency. This result suggests that wingbeat frequency likely does not limit power production in larger birds and thus cannot fully explain their inability to fly with high acceleration. Rather, muscle function may change as birds become larger, in order to compensate for the detrimental effects of body size on the amount of power produced.

5.6 Highly specialized performance: How do they do that?

It's likely that you have at some point heard or read a news story about an animal that does some task in a remarkably bizarre way. Some examples might be the scale-eating cichlid fish, which rip off and consume the scales of other fish, or the candiru, a diminutive Amazonian fish that swims up and lodges it-self in the orifices of other animals, often to the painful detriment of the hosts! Such extreme specializations often occur in a handful of species and represent examples of animals filling niches in otherwise crowded communities, such as rich tropical forests. While their morphological and functional specializa-tions may not be widespread, understanding how these animals have evolved to meet their requirements for survival can shed light on broader issues in ecology and evolution. There are numerous examples of highly specialized performance capacities that fulfill specific ecological needs. However, in most cases, the lessons are not general but species specific; therefore, we will only highlight two examples in the context of anatomical structure, namely, feeding in threadsnakes and in chameleons. More details about "extreme" perform-ance are in chapter 9.

One habitat that lends itself naturally to such specializations is the under-ground environment, particularly for larger animals, for which moving and feeding are challenging. Within such habitats, there is often a reduction in un-necessary body parts that would normally limit the ability to move in cramped circumstances. A well-documented example of reduction comes from thread-snakes (from the basal snake clade Scolecophidia), which are specialized for eating termites and ants. Burrowing snakes will sometimes specialize on ter-mites and ants, which pose certain challenges. Most termites and ants will ac-tively defend their underground lairs; consequently, many threadsnakes have smooth and overlapping scales that protect against bites from ants and ter-mites. However, most notable is the threadsnakes' extreme performance cap-acity for consuming large numbers of ants and termites (Kley and Brainerd 1999).

Most snakes consume relatively large prey, and only relatively slowly, by transporting the prey through their mouths using asynchronous ratcheting movements with their upper jaws. However, this method would be coun-terproductive for consuming relatively large numbers of ants and termites. Instead, threadsnakes exhibit a distinctive feeding mechanism in which the tooth-bearing elements within their lower jaw rotate synchronously into and out of their mouth, like a set of swinging doors, thereby pulling their prey into their esophagus. This mechanism, called "mandibular raking," is the only ver-tebrate feeding mechanism known in which prey is transported exclusively by movements of the lower jaw, and this mechanism can only occur because of the triple-jointed lower jaw present within threadsnakes. Threadsnakes pos-sess very flexible interramal and intramandibular joints, as well as more typ-ical vertebrate jaw joints. The interramal joint facilitates movement between the intramandibular joints and the tips of the mandibular rami, thereby en-abling the distal halves of the lower jaw to rotate backward into the mouth. Mandibular raking enables threadsnakes to transport prey much more rapidly than is typical for snakes, as the cycles of mandibular protraction and retraction can occur at frequencies greater than 3 Hz.

This rapid eating pattern may have arisen as a result of the threadsnakes' hazardous foraging strategy. To consume enough ant brood to survive, these snakes invade ant nests, which are vigorously defended by worker ants. Worker ants can injure or kill threadsnakes, which are small in stature (usually less than 2.5 mm in diameter and 1.5 g in weight). Across the evolution of threadsnakes, natural selection has apparently favored individuals that can feed quickly, thereby minimizing the amount of time that they are exposed to the worker ants.

Above ground, chameleons (family Chamaeleonidae) use rapid projection of their tongues to capture small prey such as insects. This specialized use of the tongue is an elaboration from other lizards, which also typically use their tongue in some manner to manipulate and capture prey. A common use of the tongue is lingual prehension, in which wet adhesion is used to capture prey, a method which works well when prey are very small. However, chameleons will also capture and eat larger prey, such as lizards or birds. Anyone who has kept a pet chameleon knows that they are not exactly fleet afoot, and thus the use of extreme tongue projection has likely evolved as an effective strategy to capture elusive prey. When propelling their tongue—the tongue retractor muscle can extend to 600% of the tongue's resting length—chameleons will create a seal on the prey item via suction. While suction typically is not long-lasting, anyone who has tried to pull a depressed plunger from a toilet knows that it can be very powerful! This mechanism allows these animals to "reel in" even large prey (Herrel et al. 2000). The tongue is propelled ballistically, which means that the only energy used to drive the tongue forward comes from the initial input, in much the same way that the flight of a bullet relies on the initial firing mechanism. Chameleons have undergone evolutionary modifications from the ancestral lizard tongue morphology to allow suction to operate via activity from two modified tongue muscles, which pull the tongue pad inwards. However, adhering to a prey item is one thing, but bringing it back to one mouth is another. Nonetheless, chameleons seem to have overcome this challenge as well: their retractor muscles have supercontractile properties that can exert powerful forces even over long tongue-projection distances. The chameleon thus represents a compelling example of how evolution of a new behavior and the concomitant morphological change has resulted in enhanced performance and the ability to use new resources.

5.7 Elastic mechanisms for powering rapid movements

Movement generally requires energy. For animals that don't possess muscles, movement can be generated through a variety of means, including hydrostatic pressure; but, for most animals, muscles facilitate locomotion via molecular motors. These include myosin, which can convert chemical energy (ATP) into potential energy. This model posits the simple principle that the speed or intensity of movement should closely correspond with the amount of energy expended. For most movements that are driven by muscle, this is likely to be true. For instance, as most mammals run faster, their energetic expenditure increases in direct proportion to speed until the animal becomes exhausted, after which muscle function declines rapidly owing to fatigue. However, the concept that increasingly fast movements should lead to increasingly higher energy expenditures is not universally true. When scientists first calculated levels of locomotor energy expenditure for mammals, they observed that, in some instances, the energy expended increased linearly with speed between low to modest speeds; however, as the animal moved at faster speeds, the energy expended increased more slowly, if at all. An example comes from jumping in kangaroos. When kangaroos are hopping slowly, their energetic expenditure increases linearly with speed; however, when they are hopping at high speeds, kangaroos expend about as much energy as if they were moving at slower speeds (Dawson and Taylor 1973). In other words, these animals seem to be gaining a large performance advantage with relatively little additional cost. Detailed investigations into the anatomy of large mammals such as kangaroos and ungulates like deer and gazelles have provided a possible reason for this energetic savings. Many large ungulates have large tendons in their limbs that act much like springs during locomotion. When their foot touches the ground during rapid locomotion, ligaments, tendons, and other soft tissues in the limb are compressed, thereby storing elastic energy like a compressed spring. As the foot leaves the ground, the pressure on the compressed soft tissue is relieved, and the elastic recoil from these spring-like structures generates additional force for propelling the animal, thus resulting in substantial energetic savings.

The fact that locomotion can be driven, at least partially, by spring-like mechanisms is one example from a broader set of ways in which elastic mechanisms can produce movement at low or nonexistent energetic costs. Elastic structures such as tendons and ligaments are important players in this game, but muscle also can exhibit some elastic capacity. The employment of elastic elements for saving energy appears to be widespread among animals and enables higher performance levels than what would be predicted based on back-of-the-envelope calculations of muscle power alone (Farley et al. 1993; Higham and Irschick 2013). Some frogs will "store" energy in tendons and ligaments just before jumping. Upon jumping, which is initiated by muscles in their hindlimbs, this stored elastic energy is then released, providing additional force that enables frogs to jump longer than would be predicted from calculations based on power outputs of their muscles (Astley and Roberts 2012). Another example is the "catapult" mechanism used by miniscule leafhopper insects when jumping, where elastic energy is stored and released through a catch mechanism (Burrows 2007). The flea is perhaps the most spectacular poster child for how elastic mechanisms can increase performance. Just before the flea jumps, resilin, an elastic protein present in its limbs, is stretched; then, the elastic energy is rapidly released, enabling the flea to jump at an extremely high acceleration. This kind of approach works well because the resilin is stretched slowly with low power input and then released quickly with high power output over a short time duration, much in the same way medieval warriors would store and release elastic energy in a catapult!

The observation that elastic elements can enable animals to store elastic energy within soft tissue and muscle during movement has been incorporated into a predictive model, otherwise known as the mass-spring model (Blickhan 1989). This model best applies to large terrestrial animals (e.g., kangaroos fig. 5.9) and envisions a heavy mass (e.g., the body) attached to a spring-like structure (e.g., a limb) that contains compressive soft tissue. During locomotion, the spring is compressed during the middle of the stance phase, and the body will then recover elastic energy at the conclusion of the stance phase. The mass-spring model is especially useful for understanding the energetic differences between walking and running. When walking, the hip follows an inverted-pendulum motion, with the hip being "vaulted" over the knee, and there is little compression of the elastic structures within the knee joint. On the other hand, when running, the hip will dip in midstance, and the elastic elements in the knee joint are thus compressed, providing some elastic storage and recoil. This difference suggests that walking in large mammals is driven primarily by muscles, while the energy for running is derived from both

Fig. 5.9 An image of a kangaroo hopping. In kangaroos, as in many large mammals, tendons and other soft tissues in the limb are stiff and thus allow kangaroos to recover elastic energy as they run. Image from Wikimedia Commons.

muscles and elastic elements, although walking does obtain some energetic savings via the manner in which the hip is first raised and then lowered, as the hip-lowering phase is energetically cheaper than the hip-raising phase because of the conversion of potential energy into kinetic energy. It is important to remember that the energetic "benefits" of spring-like structures are most apparent for larger animals and are likely less valuable for small animals such as insects. This is because larger animals can exert higher forces on soft tissues during locomotion compared to smaller animals, although, as noted above, elastic elements can enhance performance in other ways in animals such as insects and frogs.

Even within some mammal groups, there is variation in anatomical structure and locomotor behavior, and this variation can influence how much energy that can be saved by using elastic elements. Wallabies are smaller relative of kangaroos and also use hopping locomotion; however, they occur in a variety of different habitats, in some of which hopping may be more important than in others (McGowan et al. 2005, 2008). In studies comparing two wallaby species that occupy different habitats, studies of the anatomy and mechanics of locomotion has shown that the importance of spring-like elements varies among species. The tammar wallaby, *Macropus eugenii*, occupies flat open areas, whereas the yellow-footed rock wallaby, *Petrogale xanthopus*, lives on steep cliff faces. Hence, one would predict that, in tammar wallabies, hopping at relatively steady speeds for moving long distances would be ecologically relevant but that, for yellow-footed wallabies, jumping in a dynamic and nonuniform manner would likely be more important. The studies showed that, although both species use similar movements of their ankle joint during hopping, the tendons of the yellow-footed rock wallaby are generally about 13% larger than those of the tammar wallaby. However, the tammar wallaby, because of a lower mechanical advantage at the ankle joint, produced about 26% more muscular force when hopping than the yellow-footed rock wallaby did. Thus, when hopping at relatively steady and moderate speeds, tammar wallabies have a significant performance advantage over yellow-footed rock wallabies in terms of energetic savings, as they store about 73% more elastic energy in their tendons than the yellow-footed rock wallabies do. In contrast, the yellow-footed rock wallaby have tendon safety factors that are 38% higher than those for tammar wallabies. These results provide evidence that the direct match between form and function for these species was driven by ecological demands. Tammar wallabies hop steadily for long distances, so for them energetic savings are important, whereas the yellow-footed wallaby is more concerned with avoiding damage to their body elements during the often traumatic process of jumping, so the primary specialization seems to be a high safety factor, to reduce the risk of snapping a tendon when jumping off a high rock or landing awkwardly.

References

Alexander RM. 2003. Principles of Animal Locomotion. Princeton University Press, Princeton, NJ.

Astley HC, Roberts TJ. 2012. Evidence for a vertebrate catapult: Elastic energy storage in the plantaris tendon during frog jumping. Biology Letters 8:386–389.

Bernal D, Donley JM, Shadwick RE, Syme DA. 2005. Mammal-like muscles power swimming in a cold-water shark. Nature 437:1349–1352.

Blickhan R. 1989. The spring-mass model for running and hopping. Journal of Biomechanics 22:1217–1227.

Bonfil R, Meyer M, Scholl MC, Johnson R, O'Brien S, Oosthuizen H, Swanson S, Kotze D, Paterson M. 2005. Transoceanic migration, spatial dynamics, and population linkages of white sharks. Science 310:100–103.

Burrows M. 2007. Kinematics of jumping in leafhopper insects (Hemiptera, Auchenorrhyncha, Cicadellidae). Journal of Experimental Biology 210:3579–3589.

Davis RW, Polasek L, Watson R, Fuson A, Williams TM, Kanatous SB. 2004. The diving paradox: New insights into the role of the dive response in air-breathing vertebrates. Comparative Biochemistry and Physiology Part A 138:263–268.

Dawson TJ, Taylor CR. 1973. Energetic cost of locomotion in kangaroos. Nature 246:313–314.

Deban SM, Dicke U. 1999. Motor control of tongue movement during prey capture in plethodontid salamanders. Journal of Experimental Biology 202:3699–3714.

Dial KP, Biewener AA, Tobalske BW, Warrick DR. 1997. Mechanical power output of bird flight. Nature 390:67–70.

Dial KP, Green E, Irschick DJ. 2008. Allometry of behavior. Trends in Ecology and Evolution. 23:394–401.

Donley JM, Sepulveda CA, Konstantindis P, Gemballa S, Shadwick RE. 2004. Convergent evolution in mechanical design of lamnid sharks and tunas. Nature 429:61–65.

Farley CT, Glasheen J, McMahon TA. 1993. Running springs: Speed and animal size. Journal of Experimental Biology 185:71–86.

Garland T Jr. 1988. Genetic basis of activity metabolism. I. Inheritance of speed, stamina, and antipredator displays in the garter snake Thamnophis sirtalis. Evolution 42:335–350.

Garland T Jr. 1994. Quantitative genetics of locomotor behavior and physiology in a garter snake. In Boake CRB, ed. Quantitative Genetic Studies of Behavioral Evolution. University of Chicago Press, Chicago, IL, pp. 251–276.

Greer-Walker M, Pull GA. 1975. A survey of red and white muscle in marine fish. Journal of Fish Biology 7:295–300.

Herrel A, Meyers JJ, Aerts P, Nishikawa KC. 2000. The mechanics of prey prehension in chameleons. Journal of Experimental Biology 203:3255–3263.

Higham TE, Irschick DJ. 2013. Springs, steroids, and slingshots: The roles of enhancers and constraints in animal movement. Journal of Comparative Physiology B 183:583–595.

Huey RB, Hertz RB. 1984. Is a jack-of-all-temperatures a master of none? Evolution 38:441–444.

Kanatous SB, Davis RW, Watson R, Polasek, Williams TM, Mathieu-Costello O. 2002. Aerobic capacities in the skeletal muscles of Weddell seals: Key to longer dive durations? Journal of Experimental Biology 205:3601–3608.

Kley NJ, Brainerd EL. 1999. Feeding by mandibular raking in a snake. Nature 402:369–370.

Klimley PA, Le Boeuf BJ, Cantara KM, Rochert JE, Davis SF, Van Sommeran S, Kelly JT. 2001. The hunting strategy of white sharks (*Carcharodon carcharias*) near a seal colony. Marine Biology 138:617–636.

Lauder GV. 1990. Functional morphology and systematics: Studying functional patterns in an historical context. Annual Review of Ecology and Systematics 21:317–340.

Lauder GV. 1996. The argument from design. In Rose MR, Lauder GV, eds. Adaption. Academic Press, New York, NY, pp. 55–92.

McGowan CP, Baudinette RV, Biewener AA. 2005. Joint work and power associated with acceleration and deceleration in tammar wallabies (*Macropus eugenii*). Journal of Experimental Biology 208:41–53.

McGowan CP, Baudinette RV, Biewener AA. 2008. Differential design for hopping in two species of wallabies. Comparative Biochemistry and Physiology Part A 150:151–158.

Meyers JJ, Herrel A. 2005. Prey capture kinematics of ant-eating lizards. Journal of Experimental Biology 208:113–127.

Meyers JJ, Herrel A, Nishikawa KC. 2006. Morphological correlates of ant eating in horned lizards (*Phrynosoma*). Biological Journal of the Linnean Society 89:13–24.

Naples VL. 1999. Morphology, evolution and function of feeding in the giant anteater (*Myrmecophaga tridactyla*). Journal of Zoology 249:19–41.

Nishikawa K. 1999. Neuromuscular control of prey capture in frogs. Philosophical Transactions of the Royal Society of London, Biological Sciences 354:941–954.

Nishikawa K. 2000. Feeding in frogs. In Schwenk K, ed. Feeding: Form, Function and Evolution in Tetrapod Vertebrates. Academic Press, New York, NY, pp. 117–147.

Reidy SP, Kerr SR, Nelson JA. 2000. Aerobic and anaerobic swimming performance of individual Atlantic cod. Journal of Experimental Biology 203:347–357.

Rome LC, Funke RP, Alexander R McN, Lutz G, Aldridge H, Scott F. 1988. Why animals have different muscle fibre types. Nature 335:824–827.

Summers AP, Koob TJ, Brainerd EL. 1998. Stingray jaws strut their stuff. Nature 395:450–451.

Vanhooydonck B, Van Damme R, Aerts P. 2001. Speed and stamina trade-off in lacertid lizards. Evolution 55:1040–1048.

Wegner, NC, Snodgrass, OE, Dewar, H, Hyde, JR. 2015. Whole-body endothermy in a mesopelagic fish, the opah, *Lampris guttatus*. Nature 348:786–789.

Williams TM. 2001. Intermittent swimming by mammals: A strategy for increasing energetic efficiency during diving. American Zoologist 41:166–176.

6 | The evolution of performance II

Convergence, key innovations, and adaptation

6.1 How does performance evolve?

When one surveys the range of animal diversity in form and function, it's not hard to be impressed, or even confused, about the myriad of odd and remarkable phenotypes and corresponding functions. While some variation seems logical, the sheer degree of specialization is mind boggling. Examples abound of strange and fantastical phenotypes and their supposed functions. It's common for biologists to sit around and tell stories of their "favorite" adaptations! The aye-aye of Madagascar, a gracile and arboreal lemur, possesses an elongated finger for rooting out insect larvae from tree cavities. Some tropical bats possess elongated tongues of over three times their body length in order to sip nectar from flowers. The candiru catfish of the Amazon River basin has menacing gill spines that allow them to wedge themselves into body orifices or other tissues to feed on blood or other tissues. Charles Darwin himself discovered an elongated flower that clearly required an elongated tongue from some animal for pollination, a prediction that was later proven correct! Beyond the wonder of these remarkable examples, the broader point is that evolution has demonstrated great ingenuity for matching the phenotypes of animals to meet highly specialized demands. Moreover, this evolution of phenotypes has apparently co-occurred in conjunction with the evolution of functional abilities that ultimately help species to survive and prosper.

The technique of comparing different species and drawing evolutionary inferences is termed the comparative method (Felsenstein 1985; Nunn 2011). The comparative method is a powerful heuristic approach for uncovering why a trait has evolved, because it allows one to examine both the conditions that have led to the evolution of the same trait in evolutionarily different species (convergence) and the maintenance of the same trait across different species (homoplasy). In addition, the comparative approach employs rigorous statistical methods for assessing the relationship between the presence of particular traits and the presence of environmental features that are thought to be causal agents of change. However, unlike experimental methods conducted on individuals within species, the comparative method is hypothetico-deductive and therefore indirect; however, the benefit of using macroevolutionary methods is that one can examine variation at a vaster scale than in most experimental research. An instructive example of this comparative approach comes from studies of crushing ability and the morphology of jaws in vertebrates and in cartilaginous fish such as stingrays (Summers et al. 1998). Whereas most vertebrates, such as mammals, have jaws made of rigid bone that provides a robust surface for crushing hard prey, sharks and rays possess cartilage instead of bone but nonetheless are quite capable of crushing very hard prey (how would "Jaws" have become such a big hit otherwise?). For example, tiger sharks are known to consume sea turtles, and some rays will consume clams or other hard-shelled creatures, such as crabs. This remarkable feeding performance is accomplished by a simple morphological specialization. Via the calcification of the cartilage within the jaws, as well as the evolution of trabecular, or strut-like, structures within the jaws, the jaws are made far more rigid and thus far more able to crush extremely hard prey than would be expected from cartilage, especially when a "nutcracker" leverage mechanism is used. Therefore, by comparing the jaws of different kinds of vertebrates (e.g., mammals vs. rays) that accomplish the same tasks (crush hard prey) using different anatomical structures, biologists can understand how evolution molds morphology for different tasks. In addition, this comparison lends evidence to the hypothesis that the jaw structure of rays is adapted to their diet of hard prey.

However, while it is straightforward to appreciate the diversity of animal form and function and draw a simple match with environmental conditions, such an exercise does little for answering the question of how and why this diversity has evolved. In order to uncover these two aspects, the "how" and "why," it is necessary to examine performance traits using the tools of evolutionary biologists and consider these traits much the same way we might consider the evolution of a gene, for example. In this vein, an evolutionary perspective allows us to address three concepts in regards to performance capacities: (1) How have morphology and performance evolved in a similar way in unrelated taxa? (2) How do inherent constraints on morphological and performance traits guide both the evolution and the ability of animal species to

adapt to different environments? (3) What are some of the overarching selective pressures that have resulted in the evolution of animal performance traits,
and does the presence of these traits allow some animal groups to diversify to
a greater extent than for other groups?

6.2 Quantitative analysis of evolutionary data

Because the comparative method involves comparing different species, which
form a hierarchical framework (i.e., an evolutionary tree), several problems
must first be confronted. First, any comparison of distantly related taxa is confounded by numerous differences in other traits besides the one of interest.
For example, if one is interested in comparing locomotion in different species
of animals, how does one compare locomotion in reptiles versus mammals,
given their dramatic differences in evolutionary relatedness, behavior, body
size, and physiology? Whereas mammals are endothermic, for instance, reptiles are ectothermic, and this physiological difference in how they regulate
their internal body temperature influences many other aspects that could affect locomotion, such as metabolic rate, locomotor endurance, and respiratory
physiology, among others. While we cannot control for potentially confounding variables such as behavior or physiology, evolutionary biologists have developed methods to remove the confounding effects of evolutionary history.
Perhaps more importantly, such differences may not represent a confounding
factor as much as a potential explanation for differences among species.

A second problem is more subtle. A typical comparative study will examine average values for a trait across a variety of species, but an assumption of
any statistical analysis is that each data point is statistically independent, and
species means violate this assumption because any given species will be more
closely related to some species than others. Fortunately, a variety of statistical
methods are available for analyzing data from different species that corrects
for this nonindependence of data points. One commonly used method is the
independent contrasts approach pioneered by Joseph Felsenstein (1985). This
approach examines the evolutionary differences among species as the primary
unit of analysis, as opposed to raw species values. Instead of comparing the
actual values of species, we can compare differences between them, such as
the difference in bite force between species. Another way to think of this is that
every statistical analysis between species assumes evolutionary relationships
between species—it's just that conventional statistics assume that all of the
species arose from the exact same point in time and from the same ancestor
(no hierarchy). We know this isn't true, so we need to adjust the data to reflect
the hierarchical relationship between species. These new values, called contrasts, amount to "phylogeny-free" variables that we can happily use to test
evolutionary hypotheses.

6.3 A model for understanding the evolution of performance

Before we can embark on any exploration of how performance evolves, it is useful to present a general framework that offers some predictive value. If one makes the reasonable assumption that much of the variation in animal performance has arisen as a result of different environmental forces arising from different habitats, then we can hypothesize that evolution in performance capacity has arisen as a result of changes in habitat use. One can modify the basic paradigm first put forth by Steve Arnold (1983) for testing whether a trait is adaptive.

$$\text{Morphology} \rightarrow \text{Performance} \rightarrow \text{Fitness}$$

In this simple equation, morphological variation within species will be expected to result in variation in performance capacity. In turn, individual variation in performance capacity should directly result in variation in Darwinian fitness. However, this equation is primarily relevant to within-species studies, as fitness is an among-individual rather than an among-species phenomenon. Among different species, we can consider different habitats as surrogates for fitness. However, one can measure how effectively a species uses a particular habitat, thus providing a potential measure of species "fitness." If we substitute habitat use for fitness then the relevant parameters become

$$\text{Morphology} \rightarrow \text{Performance} \rightarrow \text{Habitat use}$$

Therefore, by examining all three aspects (morphology, performance, and habitat use) simultaneously, we can test whether a trait is adaptive (Garland and Losos 1994). Let's use this framework as a conceptual tool for analyzing the evolution of performance.

6.4 Convergent evolution of morphology and performance traits

One of the most basic principles of evolutionary biology is the idea that common environmental pressures should result in the evolution of similar phenotypes and functions. This process, otherwise known as convergence, such as is found in the wings of bats, birds, and insects, is some of the most powerful evidence for the process of adaptation, and this idea can also be applied to the study of performance capacities (Losos 2011). Convergence has mostly been discussed at the level of the phenotype, but this concept becomes more complex once one adds an additional layer of complexity in performance capacities. Evolutionarily different species can be convergent in morphology, but not in function, or vice versa. Indeed, the study of convergence becomes infinitely more interesting when one considers these different possibilities (Moen et al. 2013).

At the level of complex, multifaceted structures, morphological convergence has not always resulted in convergence in function (Losos 2011). Rather, evolutionarily different groups seem to have evolved very different ways of using their morphological structures to perform basic tasks that are necessary for survival. As noted, powered flight has evolved at least three times during the course of evolution in insects, birds, and bats. The ability to fly has likely enabled each of these groups to access new food resources and open new opportunities for breeding, such as via migration, among other benefits. The morphological structures used to power flight are clearly different, as birds and bats have undergone extensive evolutionary modifications in the shapes and arrangements of their forelimbs to produce a functioning wing By contrast, insects have evolved several sets of wings that are separate from their limbs which power flight. Bats employ a membranous webbing that creates a large and flexible surface area used during flight, whereas birds have evolved flexible feathers that can be lost and that can be manipulated during flight to affect flow dynamics around the wing. Finally, birds and bats are vertebrates and must generate the power for flight using their stiff bones connected via soft tissue to muscle, whereas insects are invertebrates that fly without an internal skeleton. Perhaps most importantly, these animals fly in very different ways. Birds represent a diverse group (>8,000 species) that employ flight in different ways. Some species are highly maneuverable and will typically fly only short distances, whereas other species will migrate thousands of kilometers. Birds can also fly very fast (often greater than 10 m/s [Pennycuick 1997]), and can do so for long periods of time, which means that there is a premium placed on flight economy. Bats are also highly diverse (>1,000 species), especially in terms of their diet, which includes insects, mammals, fruit, and blood! While some bats are capable of impressive long-distance movements, the fact that about 60%–70% primarily consume insects means that bats must be highly maneuverable, especially to capture elusive flying prey. Finally, insects face another challenge—namely, that they are small, ranging from less than a quarter of a millimeter to more than 15 cm, but with most falling on the smaller side. Insects use flight in a variety of different ways, but most fascinating are very small insects that are active flyers. Many aspects of insect flight in this latter group seem to defy conventional aerodynamic theory because they move their wings at such high frequencies (up to 200 Hz), fly at such unusual angles of attack, and move their wings in a variety of ways (Sane 2003), all while being able to evade a swatting hand! How each of these groups (birds, bats, and insects) have met these unique challenges of flight economy, maneuverability, and high performance at small sizes has been extensively studied by scientists using methods such as particle image velocimetry, high-speed video, and computer simulation. Researchers are only now beginning to unlock many of these animals' secrets.

Birds have evolved specialized forelimbs elaborated with feathers to achieve remarkable plasticity in their wings. Lift during flight is generated in the body, wings, and tail feathers (Tobalske 2007; Biewener 2011) and, like the membranous webbing in bat wings, wing and tail feathers will move and flex during flight, thus enabling birds to employ a range of wing, tail feather, and body movements to modulate speed and to maneuver in tight surroundings (fig. 6.1). Whereas bats employ their expanded membranes in an elastic way to generate lift and to maneuver, birds modify the direction and angle of their feathers and wings, as well as their body, during flight to accomplish the same or different behaviors.

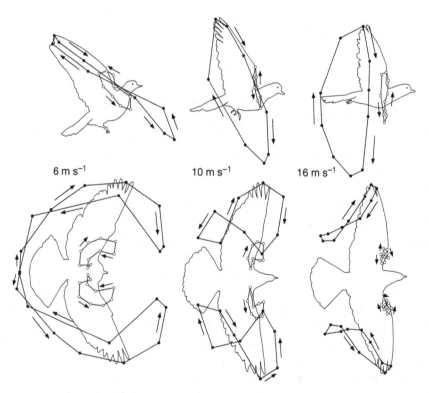

6 m s⁻¹   10 m s⁻¹   16 m s⁻¹

Fig. 6.1 A figure showing the range of motions used by birds during flight, especially as birds transition from slower to faster flight. This flexibility in wing motion may help explain how birds can fly for long distances at quite fast speeds. From Tobalske (2000) with permission from the University of Chicago press.

One of the signature features of birds is their ability to fly long distances (for some species, thousands of kilometers), often at quite high speeds. As noted in chapter 3, it's not unusual for migrating birds to obtain speeds of over 10 m/s. To put that in reference, a world-class human sprinter peaks at about 10 m/s, and that's only over 100 m! Until recently, it was widely viewed that birds had lightweight skeletons that reduced their flight load, but recent work (Dumont 2010) casts doubt on this view. Indeed, in birds, the amount of body mass derived from the skeleton is the same as in terrestrial mammals, which makes birds' long-distance feats all the more impressive. Part of their secret may lie in their ability to modulate their wing movements as they change speed. Some birds, such as pigeons, which possess "high-aspect-ratio" wings (i.e., wings that are relatively "skinny" relative to the size of the bird) transition from making relatively large three-dimensional movements with their wings to making relatively small up-and-down movements as they go faster (Tobalske and Dial 1996). Other "low-aspect" ratio birds, such as some magpies, seem to employ the same wing movement at all speeds (fig. 6.1). This ability to change wing motions may in part explain the ability of some birds to fly long distances at high speeds, although more work is ongoing in this area.

Along with alterations for some birds in how their wings are used as they fly at different speeds, wind-tunnel studies using particle image velocimetry (PIV) show different air flow patterns emanating from the wings (fig. 6.2). Just like cars, animals will often employ gaits that can minimize energy expenditure, in much the same way that traveling at 30 mph in third gear is a lot more energy efficient than traveling at the same speed in first gear! Birds appear to employ this same principle, although how these "gaits" influence the energetics of flying is still being studied. Early studies suggested that birds have two "gaits": one vortex ring for slower speeds, with additional vortices added to the wing tips at faster speeds. However, because of the complex manner by which bird wing use change with speed, it is likely that flow dynamics in wings change in a more continuous manner with speed (Tobalske 2007). The fact that flight patterns change with speed may explain an unusual pattern of energetic expenditure with speed in birds. For most animals, energetic expenditure increases approximately linearly with speed (at least until aerobic metabolism converts to anaerobic metabolism); however, in birds, this relationship seems to be "U-shaped," with high energetic expenditure at low and high speeds, and low values at intermediate speeds, although this pattern varies among species (Tobalske et al. 2003). Some of these "intermediate" speeds are quite fast, such as 5–8 m/s; thus, it is energetically cheaper for a bird to fly at 7 m/s than at 2 m/s.

Wing movements are not the only way that birds can modulate flight. Other work has shown that alterations in the angle of the body, wing, and tail during key aspects of flight (takeoff, level flight, and landing) can be achieved largely

A

B

C

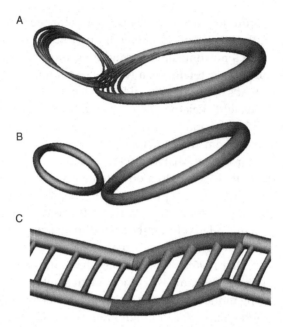

Fig. 6.2 Graphical depictions of the vortex wakes created by the wings of a thrush nightingale, *Luscinia luscinia*, at various flights speeds when in a wind tunnel. *A*, Vortex wakes created at low flight speeds. *B*, Vortex wakes created at medium flight speeds. *C*, Vortex wakes created at fast flight speeds. The vortex wakes were determined using digital particle image velocimetry. The loops on the left in the upper two panels were generated by the upstroke, as were the far left and far right "ladder" elements in the bottom panel. Taken from Tobalske (2007) with permission from the Company of Biologists Ltd.

by changes in the pitch of the body, rather than movements of the wings or tail (relative to the body; Biewener 2011). In other words, birds can modulate many aspects of flight by simply altering the position of their body, thereby potentially reducing constraints on their wings. This plasticity is coupled with great behavioral flexibility, as birds will intermix flapping with gliding, often in conjunction with prevailing winds, to minimize flight effort, and save energy (Pennycuick 2008), often through employing a form of intermittent locomotion, much in the same way that dolphins kick and glide. Altogether, this work paints a very different picture of birds than that of their being mindless flapping machines—birds are wonderfully behaviorally and functionally specialized to move their bodies in a variety of ways to enhance maneuverability and minimize energetic cost.

Bats exhibit a very different way of flying (fig. 6.3). Bat wings are an elaboration of the forelimbs, with expanded distal digits providing much of the breadth of the wing (Riskin et al. 2008, 2010). A distinctive characteristic of bat flight is the flexible nature of the wings, which are created by an expanded membrane that stretches from their modified forelimbs to their body (fig. 6.3).

Fig. 6.3 An image of a bat flying. Note the membranous tissue that creates the wing surface and which provides remarkable flexibility for flight. Image from Wikimedia Commons.

Indeed, these bat membranes are distinguished by up to 20 degrees of freedom, owing to the large number of joints in their fingers, elbows, and wrist, among others. Further flexibility in movement is provided by compliance in the bones or membranes. Kinematic studies examining the movement of various joints show the remarkable plasticity during flight in these animals (fig. 6.4). Detailed studies by Dan Riskin, Sharon Swartz, and colleagues have shown that bat wings will change in three dimensions even during a single wingbeat cycle, and that about 15 independent parameters were needed to accurately describe the wing within 95% accuracy. Another distinctive feature of bats is their kinematic flexibility, shown in a study by Iriarte-Díaz et al. (2012). By placing a load of 20% body weight on bats, the authors found that bats employed two different strategies—they either increased their flapping frequency or altered the configuration of their wings. Therefore, the three-dimensional flexibility of bat wings also enables bats to personalize their approach to a challenging functional problem.

Bat flight is also strikingly different from bird flight. Although, at first glance, the wings of bats and birds share much in common, there are also key differences. The primary feathers of birds can separate in midair so that air can flow through them, whereas the membrane in bats is a highly flexible but enclosed structure. Using the small bat species *Glossophaga soricina*, Anders

Fig. 6.4 *A–E,* Diagrams of the movements of key markers on bat wings, demonstrating the flexibility of the wings. These are ventral views, with different shadings representing different markers, such as those on the elbow, wrist, foot or tops of various digits. Taken from Riskin et al. (2008) with permission from Elsevier Press.

Hedenström and colleagues have performed elegant studies that reveal key differences between the wake vortices of bats and those of birds. The complex motion of bat wings, so amply demonstrated by Dan Riskin, Sharon Swartz, and others, results in surprising vortex loops emanating off the wings. Indeed, each bat wing appears capable of creating its own vortex loop. Further, when bats fly at faster speeds, the upstroke movement of the wing portion closer to the bat's "hand" produces negative lift whereas the portion closer to the "arm" generates positive lift (Hedenström et al. 2007). This pattern differs from that generated by birds, and the negative lift may occur because, unlike bird feathers, bats' membranous wings are not able to move apart.

In all, this wing plasticity may have played an important role in enabling bats to occupy many different ecological niches. Some bats occupy closed habitats and consequently possess relatively short, compact wings, whereas other, more wide-ranging bat species possess longer and leaner wings that enable them to fly long distances with higher efficiency (Norberg and Rayner 1987). Fast-flying bats, which have wings with a high-aspect ratio, utilize rapid movements for hawking insects, whereas longer-traveling species such as fruit bats rely on wings with a lower aspect ratio (longer and skinnier wings) that potentially compromises rapid and maneuverable flight at the expense of economical long-distance flight.

Most insects are far smaller than birds or bats, and the smallest flying insects live in a very different aerodynamic world. Insects have evolved flight in a very different manner from birds or bats—insects rely on thin and somewhat stiff wings (of which more than one pair can be employed at a time) that are thin elaborations of the insect exoskeleton. Insect wings can bend dramatically in many different directions, as in birds and bats, and present great rotational freedom over a wide range of angles (Sane 2003). As anyone who has tried to swat a fly knows, insects are capable of some pretty remarkable maneuvers that would put the US Navy's Blue Angels aviation demonstration squadron to shame. In conventional aerodynamic theory, increasing angles of attack (basically, flying at a range of angles) can lead to separation of vortices from the wings, which in a plane might lead to stall, a catastrophic loss of lift. However, small flying insects seem to thumb their noses at this rule and appear to maintain leading-edge vortices on their wings at a wide range of flight angles

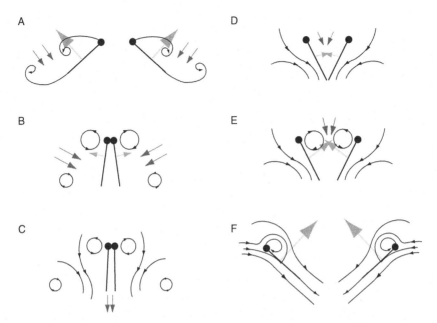

Fig. 6.5 Depictions of the complex movements of insect wings. Such movements include "clapping," shown in panels *A–C*, and "flinging apart," shown in panels *D–F*. The flow lines are shown by black lines, and the arrows depict induced velocity. The net forces acting on the airfoil are shown by arrows. This range of wing motions allows insects to defy many of conventional aerodynamic principles. Taken from Sane (2003) with permission from the Company of Biologists Ltd.

(fig. 6.5). Small flying insects throw another wrench into traditional views of flight through their complex wing-wing dynamics (Dickinson et al. 1999), which further complicate patterns of flow over wings. Overall, the evolution of these three structures (feathers, membranes, and thin outgrowths of the exoskeleton) has enabled each of these three groups to be highly successful at high-performance flight, but in different ways.

Another example of morphological convergence is provided by structures designed to enable gliding. Whereas powered flight requires the active input of energy via the wings or other structures and thus is energetically expensive, gliding employs enlarged flaps of skin or other means of increasing surface area that enable animals to gently glide from a high to a low perch. The ability to glide has evolved multiple times in animals, including in frogs, snakes, mammals, and even ants! Because gliding, unlike flapping flight, is not driven by an input of energy, the key challenge is traveling a certain distance (gliding performance), and gliding animals vary greatly in this metric, with some species, such as flying squirrels, being able to travel relatively long distances, and other species, such as flying snakes, able to travel less. The best gliders tend to possess an extensive surface area (wing) that will support their weight when gliding and also tend to be relatively small, since being large means that the force of gravity becomes ever more onerous, although being too small (e.g., like a small insect) is disadvantageous because of the ways in which body size and fluids (air) interact. Nonetheless, for entire groups of animals, therefore, becoming large poses a challenge, especially because of the relative scaling of wing area to body size. For a given change in body mass among species, one would expect that wing area should increase by only a ratio of 0.67 (McGuire 2003; McGuire and Dudley 2005). This expected ratio between a morphological trait and a metric of body size is termed isometry, which describes how morphological traits change with size without significant alterations, such as increases or decreases in proportions. The implication of this expected isometric change is that large gliding animals should have smaller ratios of wing area relative to the body mass they must support during gliding, relative to small gliding animals. Thus, one way that larger animals can maintain, or enhance their gliding capacities is through positive allometry, or the evolutionary enlargement of gliding structures beyond isometric expectations. Two aspects of flight are likely to be affected by this evolutionary scaling, namely, the velocity and the angles at which animals can glide. In turn, both variables are likely to influence how far an animal can glide. One group of gliding animals varies in size almost two-fold, namely gliding lizards (genus *Draco*; fig. 6.6), which occur in Southeast Asian forests (McGuire and Dudley 2005). Gliding may have evolved in these forests because of the relatively wide spacing among trees, as gliding may be a safer route of moving between trees than crawling on the ground would be. *Draco* lizards glide with an airfoil that is formed using winglike patagial membranes that are supported with elongated thoracic ribs. During gliding, the power required to generate aerodynamic forces is derived exclusively from the potential energy of the animal's body mass, as no oscillation of appendages, such as occurs during powered flight, is involved. However, wing area does not scale positively with body mass in these lizards but rather scales isometrically. Consequently, smaller gliding *Draco* lizard species enjoy a significant gliding advantage in gliding relative to larger *Draco* lizards and achieve higher gliding velocities and higher glide angles, thus enabling them greater flexibility for moving longer distances.

At the other end, gliding has evolved as an escape response in some animals and, consequently, these animals are less proficient gliders, as their primary goal is simply to escape the immediate area to a safe retreat (Yanoviak et al. 2005). Some ant species will descend abdomen first through very steep trajectories and at very high velocities via a method which represents a form of controlled falling (fig. 6.7). However, these ant species possess only minor evolutionary modifications for gliding relative to proficient gliders such as *Draco*. The evolution of novel behaviors also has played a key role in the evolution of gliding and showcases the interactive role of behavior and function for

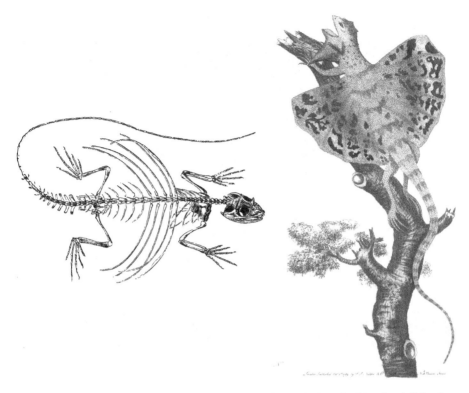

Fig. 6.6 *Left*, A skeleton of a flying lizard (*Draco* sp.). *Right*, A painting of a flying lizard. Flying lizards employ expanded membranes on specialized ribs to glide among trees. Images from Wikimedia Commons.

enhancing performance. Whereas gliding lizards or squirrels possess additional appendages that aid them during gliding, flying snakes, *Chrysopelea paradisi*, do not seem to possess any such specialized structures. Flying snakes jump from high perches and can glide to some degree by dorsoventrally flattening their body, resulting in a doubling of their body width and thus notably increasing their surface area (Socha 2002). As the snake moves more quickly during falling than when on a flat surface, the body will pitch downwards, and the head and vent will be brought toward the midpoint, thus forming an "S" shape within the horizontal plane. The snake begins to undulate laterally, starting with the anterior body. The flight trajectory shallows as lift is generated. Moreover, to sustain gliding, the snake uses an undulating form of motion unique among snakes and can thus maneuver relatively well in midair despite its lack of morphological specializations, as it can turn without banking. Indeed, there is evidence that the lateral undulation helps to generate the snake's lift.

Overall, the gliding performance of snakes, as defined by the ratio of horizontal distance gained to height lost, is about 3.7, which is notably similar to that of

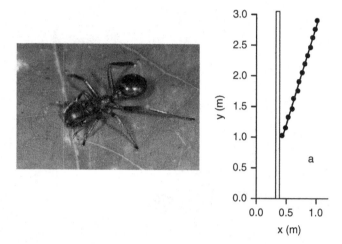

Fig. 6.7 *Left*, An image of an ant, *Cephalotes atratus*, gliding. *Right*, A typical aerial trajectory for an ant gliding onto and landing on a vertical surface. Each point indicates a 1/30 s interval. Plot taken from Yanoviak et al. (2005). Ant image is from Steve Yanoviak.

Draco lizards (3.78), although a little less than that of flying squirrels (4.77), and even better than that of flying frogs (2.19). The range of gliding performance among different animal species largely reflects their differential ecological requirements for why gliding has evolved. For those species that rely on gliding as an everyday means of moving around or eluding predators, gliding has evolved to a high level of proficiency whereas, in species in which gliding is only used as last resort to elude a startled predator, proficiency is substantially less.

6.5 Tinkering and animal performance: How variation among species has resulted in evolution of novel performance capacities

As has been noted many times, evolution has been called a tinkerer (Jacob 1977), that is, it creates complex structures, and ultimately organisms, from preexisting parts and always with the imprint of evolutionary history. Whereas an engineer can create a new part de novo that fulfills a certain need, de novo synthesis rarely happens in the process of evolution. Thus, when species evolve similar phenotypes convergently, there will be constraints and inefficiencies because of the overarching imprint of evolutionary history. How, therefore, do animals evolve performance abilities within the context of such constraints? For instance, invertebrates must move without the benefit (or hindrance?) of a stabilizing vertebral column. As another example, since spiders lack an open circulatory system, they rely largely on hydraulic pressure to move their limbs. Such differences between animals can be viewed as either advantages or disadvantages, depending on one's point of view, but we continue to be amazed at the invent-ive manner by which animals achieve remarkably high levels of performance by using very different sets of traits. Anyone watching a fruit fly understands immediately that they are far more mobile and agile than any vertebrate. Jump-ing spiders possess quite impressive jumping abilities— hence the name! In the same way that two painters with different materials will each paint a master-piece, animal groups with different anatomies can often achieve similar levels of performance. Thus, while morphology may limit performance in some cases, in other cases, the limitations simply channel evolution in another direction.

Two kinds of functional traits—suction in various marine organisms, and adhesion in insects, spiders, and lizards, stand as examples of these prin-ciples. Because both kinds of traits have evolved several times, in some cases across vastly different organisms, there are notable differences in how each has evolved. Further, within each group, there exist a variety of species with each trait, and they have all elaborated on this basic plan to suit their own ecological needs. Among fish, and some other marine animals, suction has evolved as an effective means of capturing prey. Suction is generated by the animal drawing water rapidly inward into the mouth, thus generating nega-tive pressure and thereby forcing the hapless prey item into the mouth cav-ity where it is promptly consumed (fig. 6.8). The exact definition of suction

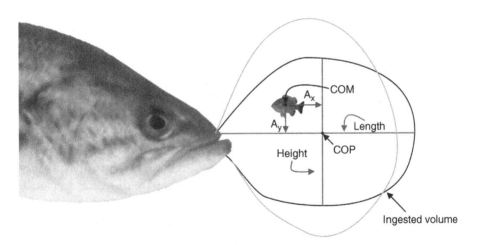

Fig. 6.8 Key elements that influence suction-feeding accuracy. The dark boundary indicates the shape of the volume of water ingested during a typical largemouth bass feeding event, and the light boundary represents that during a typical bluegill sunfish feeding event. In this system, strike accur-acy can be determined by measuring the distance from the center of mass of the prey (COM) to the center of the parcel (COP). This value can then be divided by the distance between the COP and the boundary of the ingested volume, intersecting the prey. Vertical and horizontal components (A_y and A_x, respectively) for accuracy were measured by quantifying the distance between each axis and the COM of prey. Taken from Higham et al. (2006a) with permission from the Company of Biologists Ltd.

feeding varies, but animals either draw in enough water to transport the prey entirely into the animal's mouth (inertial suction) or draw in enough water so that the animal cannot easily escape (compensatory suction). Suction feeding is widespread in fish ranging from ones that are extremely small and consume minute prey items to the spectacular giant grouper, which can swallow whole spiny lobsters (Norton and Brainerd 1993)!

While many kinds of fish can use suction to varying degrees, species that are specialized for suction typically possess several traits: a protusible upper jaw, diminished dentition, a small, laterally enclosed mouth, and strongly developed abductor muscles. The prey is drawn in or kept from moving via an explosive expansion of the buccal cavity; the resulting expansion and compression of the fluid in the cavity allows the water (and the prey) to enter. Further, pharyngeal jaws in fish may enable them to possess greater jaw flexibility and therefore superior suction performance. The ecological value of suction feeding in fish is demonstrated by the fact that there are roughly equivalent levels of suction feeding among members of different fish families; these levels have evolved independently, apparently in response to common selection pressures (Norton and Brainerd 1993).

Other animals besides fish have independently evolved suction through a variety of morphological mechanisms. For example, the ability to suction feed is also found in tadpoles of one species of frogs from the genus *Hymenochirus* (Deban and Olson 2002). The feeding mechanism for *Hymenochirus* is much like that for teleost fish, which also feed using suction through a combination of hyobranchial depression, rapid mouth protrusion, and cranial elevation, which is then followed by slower water expulsion through gill slits. Further, *Hymenochirus* and teleosts also share a round mouth opening, which is hydrodynamically advantageous for suction feeding. Seahorses can also feed by suction (Van Wassenbergh et al. 2009); they rapidly rotate their head close to the prey and use rapid and powerful suction powered by elastic recoil in tendons for the muscles that are responsible for head rotation (i.e., levator neurocranialis, which is a derivative of the epaxial musculature; see fig. 6.9).

The beauty of suction feeding is that one can straightforwardly quantify how well an animal can "suck" and therefore compare different species in terms of their suction feeding performance. Suction performance can also be quantified by measuring the amount of negative pressure generated within the buccal cavity of fish, as pressure levels correlate with the speed of water entering the mouth during the suction event. Some fish species are outstanding suction feeders and can draw in objects very quickly, whereas other species are far less proficient, but can ingest a large volume of water (Higham et al. 2006a, b). There are several primary determinants of suction feeding in fish. One factor is the size of epaxial musculature in the head. If fish have large amounts of muscle, that allows them to create rapid kinematic movements and therefore large negative changes in buccal pressures and high

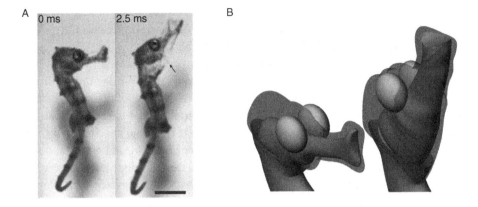

Fig. 6.9 *A,* An image of the different movements of the head performed by sea horses during suction feeding. *B,* Degrees of rotation of the head and feeding apparatus during suction feeding. Taken from Van Wassenbergh et al. (2009) with permission from Royal Society Publishing.

suction velocities (Carroll et al. 2004). Fish species also vary in how protusible their jaws are and, for those species that can protrude their jaws a significant amount (up to 35%), total suction forces are high. This factor may explain the evolution of extreme jaw protusion in fish, as this feature may be an adaptive mechanism for enhancing performance.

Research on different species of centrarchid fishes reveals that overall mouth size is a key determinant of suction performance. As larger fish can generate larger suction forces, then size matters a lot! However, among fish of similar size, there exists variation in the relative size of the mouth. Fish with small mouths often specialize on small and elusive prey, in many cases using their jaws as a pair of delicate and rapid tweezers, and therefore this reveals a potential trade-off between the ability to target and capture elusive prey, and the ability to consume larger prey via suction. A second factor is behavioral—fish that can rapidly open their mouths and achieve a peak gape size can generate higher pressures than species that perform this behavior more slowly. Fishes also employ different strategies, which can impact their ability to access prey through suction. In studies using digital PIV, a method that utilizes lasers to visualize the motion of the fluid, it was shown that bluegill sunfish, *Lepomis macrochirus,* are able to generate higher suction forces than largemouth bass, *Micropterus salmoides,* which ingest a greater volume of water and use more ram (swimming) for capturing prey (Higham et al. 2006a, b). This work suggests that suction alone is part of a suite of traits used by fish to capture prey. The fact that suction feeding requires very accurate strikes, owing to the fact that the flow field generated only influences a small region outside the mouth, means that the locomotor system is critical for positioning the mouth appropriately. Almost all fish must swim up to a prey before attempting to capture it, and this observation reflects the precise integration between complex systems (Higham 2007).

The notion of evolution as a tinkerer is nicely demonstrated by the evolution of adhesive capacities in insects, spiders, and lizards. While many animals can climb, these three groups are unique because they have all evolved some form of dry adhesion, or the ability to adhere to surfaces without the aid of glue-like secretions (although secretions are used by some species). All three groups have evolved setae, minute elaborations of the dermal layer that occur together in dense clumps, much like a forest of trees (Ruibal and Ernst 1965; Russell 1975; see fig. 6.10) These groups use then to some extent for different tasks, such as climbing in geckos, grasping prey in spiders and kissing bugs, and holding onto mates in lady bugs. Setae differ in size and shape, ranging from relatively large and single-tipped setae on the feet of kissing bugs to the far smaller and multibranched stalks of geckos. However, while pad-bearing lizards (fig. 6.11) have all evolved setae, insects and spiders have evolved several other kinds of structures that aid in climbing. Whereas the pads of some insects (e.g., beetles) are covered with setae, other insects, such as cockroaches, possess soft deformable pads that have a smooth (nonhairy) surface (Arzt et al. 2003). Moreover, some insect species have evolved tiny claw-like appendages that enable them to cling to rough surfaces, but these other structures come into play when the need arises for adhering to smoother surfaces.

The morphology of adhesive structures, and their resultant performance capacities, shows a strong link with the occupation of different habitats or the use of different behaviors (Irschick et al. 2006). Among various species of *Anolis* lizards, the size of the toepad (relative to body size) and the ability to cling to smooth surfaces shows a close evolutionary match to the average height of the canopy at which they occur (Elstrott and Irschick 2004). This means that evolutionary pressures have likely favored higher clinging capacities to enable these animals to climb proficiently in ever higher places. Ants and plants

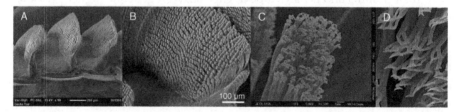

Fig. 6.10 Foot structures from the Tokay gecko, *Gekko gecko. A*, Lamellae. *B*, Setal arrays. *C*, Setal clumps. *D*, Individual setae. These structures, along with the internal anatomy of gecko feet, allow geckos to climb vertical and overhanging surfaces. Images from Michael Bartlett, courtesy of the University of Massachusetts at Amherst.

Fig. 6.11 An image of a toepad from the gecko *Gehyra vorax*. Elaborated toepads and toepad substructures such as setae and soft tissue (tendons) enable these animals to climb surfaces. Image by Duncan J. Irschick.

sometimes form such symbiotic relationships: for example, although plants in the genus *Macaranga* possess slippery epicuticular wax crystals that normally prevent insects from running on the plant surface, a group of "waxrunner" ants (members of the genera *Crematogaster, Technomyrmex*, and *Camponotus*) live in association among these waxy plants (Federle et al. 2000). These ants appear to have made a trade-off between the ability to run on more standard smooth surfaces and the capacity to run on waxy surfaces.

6.6 Key innovations, adaptive radiation, and the invasion of novel habitats

Some animal groups are more abundant than others. While there are thousands of species of bacteria, there are far fewer bear species. Moreover, this differential diversity is particularly notable when one closely examines the composition of communities. Within most ecological communities, there often occur many species from certain groups, while other groups are rare. Evolutionary analyses allow us to ask why some animal groups have apparently prospered in some habitats or regions whereas others have not. This pattern of differential abundance of species within ecological communities is driven by many factors, but adaptive radiation is one process that is known to result in an explosion of species that often dominate ecological communities. Adaptive radiation refers to the rapid explosion of phenotypes driven by intense competition for access for different resources (Schluter 2000). In many cases, the evolution of performance capacities plays a key role in facilitating access to these resources.

Adaptive radiations can occur in various ways, but one factor that is thought to facilitate this process is the evolution of key innovations. Key innovations are morphological traits or properties that act as a spur to animal diversity, usually because they allow animals to access previously inaccessible resources or habitats or because they allow animals to more rapidly evolve novel morphological traits and functional capacities. Key innovations can involve enhanced performance capacities that are a critical ingredient in their success. For example, coral reef fishes are well known for their high diversity, and labrid fish (family Labridae) are among the most diverse and prominent members. Much of their diversity can be traced to their seemingly limitless range of jaw types that have allowed them to consume a wide variety of different food types ranging from mollusks to algae to other fish (fig. 6.12). In this fashion, these fish have diversified into many different niches and have evolved many different morphological and behavioral specializations for different food types (fig. 6.13). Part of the secret to this diversity is that labrid fishes display a pattern of divergence within geographic regions and

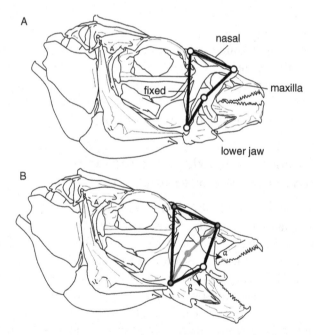

Fig. 6.12 Depictions of the oral jaws for a labrid fish demonstrating the four-bar linkage. *A*, Jaws when open. *B*, Jaws when closed. These four bone elements (fixed, lower jaw, nasal, and maxilla) are connected at various mobile joints to generate a loop. An emergent mechanical property of this system, namely, the maxillary kinematic transmission coefficient (maxillary KT = α/β), defines the degree of rotation of the jaw. The flexible nature of labrid fish jaws enables them to access a wide range of prey items. Taken from Alfaro et al. (2005) with permission from the University of Chicago Press.

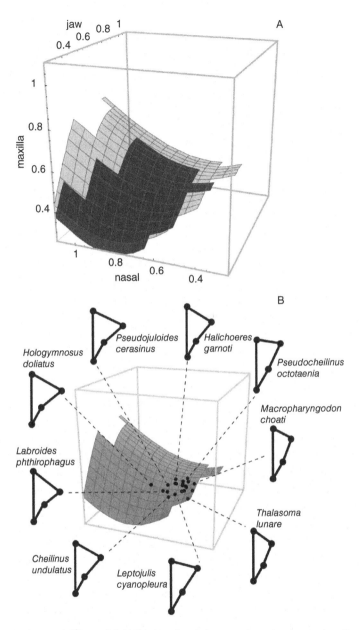

Fig. 6.13 The range of different labrid fish functional phenotypes can be shown via a three-way plot of lower jaw, maxilla, and nasal links, which are all expressed in terms of a proportion of a fixed link. *A*, The range of maxillary kinematic transmission (KT) coefficients as a function of these three variables. *B*, KT values for different labrid species. Even though they occupy a relatively small range in possible morphospace, labrid fishes with similar maxillary KT values demonstrate a high degree of diversity. This functional flexibility allows these fish to exhibit many-to-one mapping in which a limited range of phenotypes can access a wide range of food types. Redrawn from Alfaro et al. (2005) with permission from the University of Chicago Press.

convergence among them in the way their jaw has been used. One kind of jaw is well suited for rapid movements, and such mechanically fast jaws have arisen at least 14 times, in all cases evolving from ancestors that have more powerful jaws. In other words, evolutionary flexibility in the jaw and in the relationship between morphology and performance certainly is one of the keys to producing a diverse lineage.

While fish jaws vary in many ways, one important innovation is the pharyngeal jaw, which may have allowed groups that possessed it a greater flexibility to evolve a wide range of jaw phenotypes and therefore to consume a variety of food items (Liem 1973), although that view is still debated (Alfaro et al. 2009). A well-studied organism that may have benefited from the evolution of this trait are labrid fishes, of which over 600 species exist. These fish occupy a wide range of habitats, from tropical to temperate marine ecosystems, and are among the most conspicuous members of coral reefs around the world. The pharyngeal jaw represents a "second" jaw that resides in the back of the mouth and aids in processing seized prey and thus potentially enables the front jaws to expand into a variety of many other roles, in essence freeing them from having to serve the dual roles of seizing prey and processing them. This integrated and specialized key innovation allows cichlids to not only transport (deglutination) but also to prepare food, thereby freeing the premaxillary and mandibular jaws to evolve other specializations for consuming diverse kinds of food. The pharyngeal jaw is thus a compelling example of how the evolution of morphological novelties can open the door to new functions, although the link between the origin of this trait and increased diversification remains unclear. There are a myriad of other examples of how functional traits have opened the door to new habitats, and may have facilitated adaptive radiations. These include the evolution of toepads in lizards (anole lizards, geckos, and skinks), and the evolution of beak shape and size in Darwin's finches in the Galapagos, as well as in Hawaiian honeycreepers (Schluter 2000).

Looking beyond the concept of adaptive radiations, there is also the problem of how animals can occupy novel habitats that possess key challenges, such as the presence of dangerous predators. Overcoming this challenge often requires the evolution of enhanced performance capacities that allows animals to elude predators. In the Northern Lake region of the USA, damselflies, genus *Enallagma*, have apparently solved the problem of invading lakes full of novel predators by developing more rapid burst speeds via biochemical evolution (McPeek et al. 1996; McPeek 1999). When damselflies from lakes in which fish were the top predators moved to lakes where the top predators were dragonfly larvae, this ecological transition posed new risks for them. In order to effectively elude dragonfly larvae, damselfly larvae must use short bursts of locomotion that would normally be suicidal in the presence of a much larger and faster fish predator. Thus, damselflies in lakes with dragonfly larvae as the top predators have evolved burst speeds that are faster than those of damselflies in lakes where fish are the top predators.

The evolution of this enhanced capacity in damselflies was driven by alterations in physiology. Enzymes dictate many performance traits because enzymes are instrumental in fueling both anaerobic and aerobic pathways that are used during dynamic movements such as sprinting, feeding, or vocalization. Many aspects of morphology and physiology influence burst speed; however, in this case, three enzymes play particularly important roles: pyruvate kinase, lactate dehydrogenase (both used in glycolysis), and arginine kinase (involved in reestablishing the pool of available ATP). Fitting a priori predictions, damselflies that have colonized dragonfly lakes show significantly greater activity for arginine kinase but not for the two other enzymes. Because these damselflies move only in very short bursts, the increased activity of arginine kinase may extend the period of maximal exertion for several seconds, providing a crucial window of opportunity for these animals to effectively escape. Although indirect, this example is evidence that variation among species in arginine kinase activity is adaptive because of its role in influencing the increased burst speeds that have arisen as a consequence of invasion of habitats with novel predators.

Just as occupying habitats with dangerous predators often requires the evolution of new specializations, the occupation of extreme environments also poses challenges. The occupation of different elevational zones is a case in point, as the partial pressure of oxygen becomes much lower as elevation increases, so that less is available for usage by animals (hypobaric hypoxia). Moreover, higher elevations also present lower air densities and ambient air temperatures, both of which can affect flight. For animals with high oxygen demands, this is a significant challenge. For hummingbirds, which occur across a wide range of elevations from sea level to about 4,000 m in the South American Andes, there is some evidence that this invasion of habitats has been progressive (Altshuler and Dudley 2002). In hummingbirds, flight is highly energetically expensive because of their extensive use of rapid wing movements, which can exceed 50 Hz in frequency. Evolutionary analyses show that hummingbirds have achieved larger sizes and relatively larger wing areas as they have invaded higher elevations. Hummingbirds have also undergone behavioral changes in the way that they move their wings in terms of wingbeat kinematics and amplitude. Hummingbirds have thus met the challenge of occupying higher elevations by using greater lift power reserves, along with a poorly understood insensitivity to hypoxia. Further, kinematic adjustments, such as increasing stroke amplitude, enable these animals to move effectively in the hypodense air typically found at high elevations, whereas the energetically more expensive strategy of increasing wingbeat frequency is rarely used. Overall, evolutionary modifications of behavior appear to be the primary mechanism that has enabled these animals to specialize for conditions at higher elevations.

References

Alfaro ME, Bolnick DI, Wainwright PC. 2005. Evolutionary consequences of a redundant map of morphology to mechanics: An example using the jaws of labrid fishes. American Naturalist 165:E140–E154.

Alfaro ME, Brock CD, Banbury BL, Wainwright PC. 2009. Does evolutionary innovation in pharyngeal jaws lead to rapid lineage diversification in labrid fishes? BMC Evolutionary Biology 9:255.

Altshuler DL, Dudley R. 2002. The ecological and evolutionary interface of hummingbird flight physiology. Journal of Experimental Biology 205:2325–2336.

Arnold SJ. 1983. Morphology, performance, and fitness. American Zoologist 23:347–361.

Arzt E, Gorb S, Spolenak R. 2003. From micro and nano contacts in biological attachment devices. Proceedings of the National Academy of Sciences 100:10603–10606.

Biewener A. 2011. Muscle function in avian flight: Achieving power and control. Proceedings of the Royal Society of London B 366:1496–1506.

Carroll AM, Wainwright PC, Huskey SH, Collar DC, Turingan RG. 2004. Morphology predicts suction feeding performance in centrarchid fishes. Journal of Experimental Biology 207:3873–3881.

Deban SM, Olson WM. 2002. Suction feeding by a tiny predatory tadpole. Nature 420:41–42.

Dickinson MH, Lehmann FO, Sane SP. 1999. Wing rotation and the aerodynamic basis of insect flight. Science 284:1954–1960.

Dumont ER. 2010. Bone density and the lightweight skeletons of birds. Proceedings of the Royal Society of London B. 277:2193–2198.

Elstrott J, Irschick DJ. 2004. Evolutionary correlations among morphology, habitat use and clinging performance in Caribbean *Anolis* lizards. Biological Journal of the Linnean Society 83:389–398.

Federle W, Rohrseitz K, Holldobler B. 2000. Attachment forces of ants measured with a centrifuge: Better "wax-runners" have a poorer attachment to a smooth surface. Journal of Experimental Biology 203:505–512.

Felsenstein J. 1985. Phylogenies and the comparative method. American Naturalist 125:1–15.

Garland T Jr, Losos JB. 1994. Ecological morphology of locomotor performance in squamate reptiles. In Wainwright P, Reilly SM, eds. Ecological Morphology: Integrative Organismal Biology. University of Chicago Press, Chicago, IL, pp. 240–302.

Hedenström A, Johansson LC, Wolf M, Busse R von, Winter Y, Spedding GR. 2007. Bat flight generates complex aerodynamic tracks. Science 316:894–897.

Higham TE. 2007. The integration of locomotion and prey capture in vertebrates: Morphology, behavior, and performance. Integrative and Comparative Biology 47:82–95.

Higham TE, Day SW, Wainwright PC. 2006a. Multidimensional analysis of suction feeding performance in fishes: Fluid speed, acceleration, strike accuracy and the ingested volume of water. Journal of Experimental Biology 209:2713–2725.

Higham TE, Day SW, Wainwright PC. 2006b. The pressures of suction feeding: The relation between buccal pressure and induced fluid speed in centrarchid fishes. Journal of Experimental Biology 209:3281–3287.

Iriarte-Díaz J, Riskin DK, Breuer KS, Swartz SM. 2012. Kinematic plasticity during flight in fruit bats: Individual variability in response to loading. PLoS ONE 7:e36665.

Irschick DJ, Herrel A, Vanhooydonck B. 2006. Whole-organism studies of adhesion in pad-bearing lizards: Creative evolutionary solutions to functional problems. Journal of Comparative Physiology A. 192:1169–1177.

Jacob F. 1977. Evolution and tinkering. Science 196:1161–1166.

Liem KL. 1973. Evolutionary strategies and morphological innovations: Cichlid pharyngeal jaws. Systematic Zoology 22:425–441.

Losos JB. 2011. Convergence, adaption, and constraint. Evolution 65:1827–1840.

McGuire JA. 2003. Allometric prediction of locomotor performance: An example from Southeast Asian flying lizards. American Naturalist 161:337–349.

McGuire JA, Dudley R. 2005. The cost of living large: Comparative gliding performance in flying lizards (Agamidae: *Draco*). American Naturalist 166:93–106.

McPeek MA. 1999. Biochemical evolution associated with antipredator adaptation in damselflies. Evolution 53:1835–1845.

McPeek MA, Schrot AK, Brown JM. 1996. Adaptation to predators in a new community: Swimming performance and predator avoidance in damselflies. Ecology 77:617–629.

Moen DS, Irschick DJ, Wiens JJ. 2013. Evolutionary conservatism and convergence both lead to striking similarity in ecology, morphology and performance across continents in frogs. Proceedings of the Royal Society of London B 280: 20132156.

Norberg UM, Rayner JMV. 1987. Ecological morphology and flight in bats (Mammalia; Chiroptera): Wing adaptation, flight performance, foraging strategy and echolocation. Philosophical Transactions of the Royal Society of London, Biological Sciences 316:335–427.

Norton SF, Brainerd EL. 1993. Convergence in the feeding mechanics of ecomorphologically similar species in the Centrarchidae and Cichlidae. Journal of Experimental Biology 176:11–29.

Nunn CL. 2011. The Comparative Method in Evolutionary Anthropology and Biology. University of Chicago Press, Chicago, IL.

Pennycuick CJ. 1997. Actual and "optimum" flight speeds: Field data reassessed. Journal of Experimental Biology 200:2355–2361.

Pennycuick CJ. 2008. Modelling the Flying Bird. Elsevier Press, Amsterdam.

Riskin DK, Iriarte-Diaz J, Middleton KM, Breuer KS, Swartz SM. 2010. The effect of body size on the wing movements of pteropodid bats, with insights into thrust and lift production. Journal of Experimental Biology 213:4110–4122.

Riskin DK, Willis DJ, Iriarte-Diaz J, Hedrick TL, Kostandov M, Chen J, Laidlawd DH, Breuer KS, Swartz SM. 2008. Quantifying the complexity of bat wing kinematics. Journal of Theoretical Biology 254:604–615.

Ruibal R, Ernst V. 1965. The structure of digital setae of lizards. Journal of Morphology 117:271–294.

Russell AP. 1975. A contribution to the functional analysis of the foot of the Tokay, Gekko gecko (Reptilia: Gekkonidae). Journal of Zoology London 176:437–476.

Sane SP. 2003. The aerodynamics of insect flight. Journal of Experimental Biology 206:4191–4208.

Schluter D. 2000. The Ecology of Adaptive Radiation. Oxford University Press, Oxford.

Socha JJ. 2002. Kinematics: Gliding flight in the paradise tree snake. Nature 418:603–604.

Summers AP, Koob TJ, Brainerd EL. 1998. Stingray jaws strut their stuff. Nature 395:450–451.

Tobalske BW. 2000. Biomechanics and physiology of gait selection in flying birds. Physiological and Biochemical Zoology 73:736–750.

Tobalske BW. 2007. Biomechanics of bird flight. Journal of Experimental Biology 210:3135–3146.

Tobalske BW, Dial KP. 1996. Flight kinematics of black-billed magpies and pigeons over a wide range of speeds. Journal of Experimental Biology 199:263–280.

Tobalske BW, Hedrick TL, Dial KP, Biewener AA. 2003. Comparative power curves in bird flight. Nature 421:363–366.

Van Wassenbergh S, Roos G, Genbrugge A, Leysen H, Aerts P, Adriaens D, Herrel A. 2009. Suction is kid's play: Extremely fast suction in newborn seahorses. Biology Letters 4:200–203.

Yanoviak SP, Dudley R, Kaspari M. 2005. Directed aerial descent in canopy ants. Nature 433:624–626.

7 | Trade-offs and constraints on performance

7.1 Trade-offs, performance, and optimization

Have you ever wondered why you cannot run as fast as a cheetah or fly as fast as a peregrine falcon? And why can't cheetahs run even faster? A casual glance at the panorama of animal diversity reveals a surprising number of limitations that would seem to be counterintuitive in the context of evolution. Indeed, if we compare the human body relative to an engineer's view of how humans should work, we fall far short (at least the authors do!). The simple answer to this conundrum is that evolution does not always proceed to what is the best solution but rather to which solution is possible given a suite of constraints. Animals can be really bad at something relative to some theoretical optimum, but if it's enough to survive and reproduce, then that trait should be favored by natural selection. This view shifts the debate from whether a trait or performance ability is perfect (relative to a theoretical optimum) to whether it is simply adapted to its surroundings. Early views of the evolution of the phenotype and function emphasized how traits were extremely well adapted to their environments and even led to some discussion of how all aspects of the phenotype had some function. However, over the past 25 or so years, there has been a considerable tempering of this view, and there is now a greater appreciation that evolution operates in the context of constraints and trade-offs. Steven Jay Gould and Richard Lewontin wrote an influential paper in 1979 that explained how the designs of certain animal (human and nonhuman) structures have likely arisen because of constraints inherent in the system (Gould and Lewontin 1979). These "constraints" take on many forms. Developmental constraints can limit the range of available phenotypes that are accessible to natural selection (ever wondered why we have five fingers and not seven?). Some have argued that the simple pattern of ancestry can impose constraints, as

Animal Athletes. Duncan J. Irschick and Timothy E. Higham. © Duncan J. Irschick and Timothy E. Higham 2016. Published in 2016 by Oxford University Press.

close relatives are often similar to one another in the evolutionary tree of life. These constraints have limited and channeled diversity for millions of years and, in some cases, may have prevented animals from evolving structures or abilities. Constraints can also be based on basic physical laws about the way the world works. For example, muscle attaches to bone via tendons, and the size of the muscle cannot exceed a certain value for a given size and shape of a bone. If the muscle is too large, the bone will simply break when force is exerted on it, or the tendon might come off the bone. This limitation, or constraint, is important to consider when thinking about how animals look and why they do the things they do (Biewener 1990). Finally, perhaps the primary reason for why we should not expect any structure or ability to be optimal is that the body works together as a unit, and there are trade-offs between the ability to perform well at one task, and the ability to perform well at others. In short, there are many reasons for why we cannot run as fast as cheetahs, or fly as fast as birds, and let's explore these ideas further.

7.2 Why expect a trade-off?

There are also several reasons for why some morphological traits and performance capacities cannot be optimized without sacrificing some other feature. First, because the production and maintenance of morphological structures requires energy, especially during development, investment in one structure typically occurs at the expense of other traits (Lailvaux and Husak 2014). This occurs for three main reasons: limited energy budgets, physical space constraints, and mechanical constraints. Unlike many modern humans, most wild animals operate on a food tightrope and rarely have access to excess food beyond their minimal daily nutritional requirements. Further, real estate within the body is expensive in living animals. There is a limited amount of space for the placement and size of morphological structures such as muscles, nerves, soft tissues, or other organs. For example, bones and tendons support only a certain mass of muscle and, if a particular muscle group contains large numbers of fast-twitch fibers, there is less room for slow-twitch fibers. When this concept is violated, such as when human athletes abuse steroids and thereby increase overall muscle mass beyond a normal size range, then catastrophic injuries can occur, such as torn tendons. A final constraint on functional systems concerns inherent mechanical trade-offs within functional systems. For example, the maximum speed of locomotion of an organism is largely a function of two factors: stride frequency and the power output of individual muscles. Maximum power output requires, in addition to velocity, increases in the maximum force that a muscle can produce, and maximum force is a function of muscle cross-sectional area and the underlying properties of the muscle itself. Hence, a muscle composed largely of fast-twitch muscle fibers is capable of producing large amounts of power but is also easily exhausted (fig. 7.1). Conversely, a muscle that is composed largely of slow-twitch muscle fibers is not capable of producing as much power but is less easily fatigued. The irreconcilable nature of these factors at the level of muscle fibers means that, for the whole organism, there are also likely to be trade-offs between the production of force (i.e., strength) and endurance, suggesting that it should be challenging for an individual or a species both to be very fast and to possess high endurance capacities.

Altogether, these three factors—energetic limitations, space limitations, and mechanical trade-offs—lead to the a priori expectation that many performance capacities can be optimized only at the expense of other performance capacities. However, as you have seen from other examples in this book, animals have evolved some ingenious means of overcoming such limitations! One implication of these trade-offs is the idea that animals cannot have their cake and eat it too. A more fancy way of stating this is the

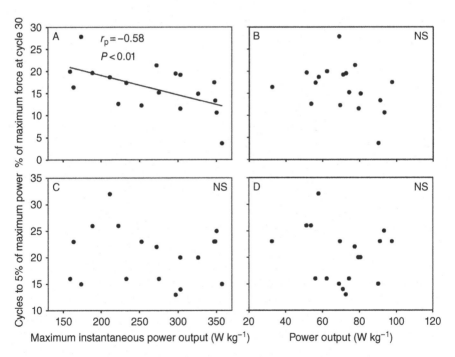

Fig. 7.1 A–D, Relationships between maximum power output and fatigue resistance for the peroneus muscle in *Xenopus laevis*. Note that "Cycles to 5% of maximum power" indicates the number of cycles that it took to reach 5% of the initial value at cycle 1. Graph taken from Wilson et al. (2002) with permission from the Company of Biologists Ltd.

"Jack-of-all-trades-master-of-none" hypothesis, which states that an organism or species that performs many tasks reasonably well is unlikely to perform any of them particularly well. Conversely, extreme specialists are likely to be extremely good performers at a particular task but are likely to be spectacularly bad at other tasks. If we accept this idea, which is still being tested and evaluated, then we can ask how individual animals and species fall on this spectrum of different possible levels of performance specialization.

7.3 Trade-offs within and among species

Trade-offs can occur at different levels, such as among individuals within a species or among well-defined species. For the among-species comparison, an analogy would be comparing baseball players to football players at some common task. By contrast, an among-individual analogy would be comparing different individual football players. Within a species, the degree of variation in morphology and performance capacity is typically far less than that among species. For example, if one compares a species specialized for speed, such as the cheetah, to one specialized for endurance, such as a species of wild dog, one would notice many morphological differences that enable these animals to move in different ways (Hudson et al. 2011). By contrast, among individuals, the degree of variation in morphology and performance is subtler, and trade-offs may be harder to detect. Indeed, it is not unusual to detect trade-offs among but not within species. One interesting exception is the killifish, *Rivulus hartii*, from Trinidad; killifish populations with high levels of predation exhibit greater sprint speeds (for escaping predators) but decreased stamina at the critical swimming speed (Oufiero et al. 2011).

Differences in trade-offs among and within species can be observed within the context of the distribution and effects of red and white muscle, and the consequent effects of this distribution on locomotion and activity levels. As noted in chapter 5, the roles of red and white muscle in different kinds of performance have been well studied. Species that rely on extended bouts of movement often possess relatively large amounts of red muscle that fuel their impressive aerobic capacities (Greer-Walker and Pull 1975; Rome et al. 1988). For species that rely on quick bursts of speed to capture prey, there is a greater specialization toward anaerobic locomotion fueled by relatively large amounts of white muscle. For example, some pelagic fish species, such as tuna and some sharks, will migrate long distances at relatively fast speeds and therefore possess large amounts of red muscle (fig. 7.2), whereas more sedentary fish species tend to possess less red muscle, and more white muscle (Greer-Walker and Pull 1975). However, many animals often require capacities for high endurance and rapid movements and have a complex mosaic of red and white muscle that can be differentially recruited to fulfill these ecological needs.

Fig. 7.2 Image of red muscle (dark circular areas near the midline) from a cross-section of a porbeagle shark, *Lamna nasus*. Image is from Wikimedia Commons.

Given the different requirements of white and red muscle, and their different roles for aerobic and anaerobic locomotor performance, one might expect a trade-off between the ability to run at high speeds and the ability to run with high endurance. Among lacertid lizard species, there is a trade-off between maximum speed and endurance, such that fast lizard species have poor endurance, and vice versa (Vanhooydonck et al. 2001; see fig. 7.3). However, when different individuals are compared, such as within humans or within garter snakes, there is little evidence for a trade-off between maximum speed and endurance (Garland 1988, 1994), with some exceptions, such as in cod, *Gadus morhua* (Reidy et al. 2000). One explanation for these conflicting results is that anatomical differences among individuals are modest compared to the myriad of other factors, such as motivation or other physiological variables that affect performance traits such as maximum speed and endurance. Among divergent species that differ in the relative proportions of muscle fibers and even their anatomical structure, variation in performance due to other factors may be inconsequential compared to these overarching factors. More studies linking the exact muscle physiology and performance across individuals and species might resolve these conflicting results.

7.4 Mechanical constraints on performance

Evolution has been sparing in its paintbrushes, and animals must often perform multiple tasks with the same structure. Bird song is a case in point. Birds

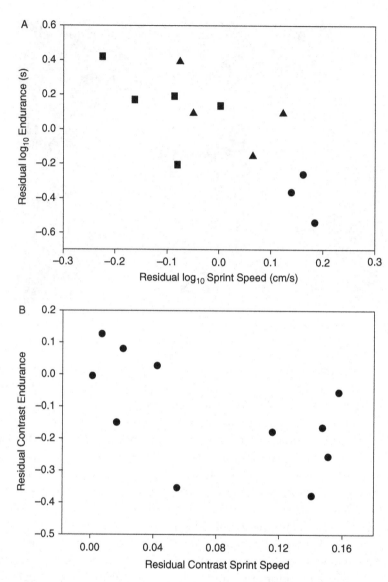

Fig. 7.3 The relationships between sprint speed and endurance among 12 species of lacertid lizards. *A*, The log-transformed residuals from the means for each species; circles represent ground-dwelling species in open habitats, squares represent ground-dwelling species in vegetated habitats, and triangles represent rock-dwelling species. *B*, The residuals of the phylogenetic contrasts through the origin. Graph taken from Vanhooydonck et al. (2001) with permission from Wiley-Blackwell Press.

sing with their beaks, but they must also feed. For birds that eat a variety of prey items and for which food is reasonably abundant, this trade-off may be easy to overcome; however, for species that consume specialized kinds of prey and for which food is sometimes scarce, this trade-off takes on a greater importance. As noted in chapter 1, not all kinds of bird song clearly fit the criterion of a performance trait as defined for this book, but the underlying principles of this example are compelling enough that it merits some discussion. The finches of the Galapagos Islands have evolved particular beak shapes that strongly influence both the size and the hardness of the seed types that they can consume (Grant and Grant 2014). Some species have large and powerful beaks that can crack large and hard seeds, whereas other species have smaller and more slender beaks that are used for manipulating and consuming smaller and less robust seeds.

This specialization to different seed types appears to have influenced the evolution of song. Sound production in birds occurs through a complex of interrelated structures. The syrinx is the primary sound producing structure in birds and can be analogized to the human larynx. Air flows from the lungs into the syrinx, causing these tissues to vibrate, thus producing sound as air is blown out through the beak. Other structures, such as the medial tympaniform membranes, are thought to play a role in singing in songbirds. Of course, the syrinx does not act alone but instead acts in concert with a variety of other structures to produce sound. Birds control their sounds by opening and closing their beaks, effectively lengthening or shortening their vocal tracts, respectively, and thus changing the resonance of the vocal tract. A prediction stemming from this process is that birds will open their beaks more widely when making higher-pitched sounds than when making lower-pitched sounds.

Because of the array of beak types among different species of Darwin's finches, we can ask whether there is any impact of break size and shape on song. For example, have bird species with large beaks, and thus larger vocal tracts, evolved songs that exhibit lower vocal frequencies? Another factor is beak shape. Finch species with powerful beaks that are well suited for crushing seeds may be constrained in their singing ability because of a trade-off between force and velocity (Herrel et al. 2009), whereas species with more gracile beaks may be less constrained. Indeed, beak gape correlates with vocal frequency range for all finch species that were studied. Further, the trend for species to match beak gape to source frequencies appears to be ancestral, as it has been conserved during the course of finch radiation. These results suggest that specialization for feeding occurred first, with concomitant effects on singing occurring secondarily. Comparative studies also suggest constraints on singing performance across bird species. Among 34 bird species, including several finch species, two acoustic variables—trill rate and frequency bandwidth—stand out as being especially important (Podos 1997). Trills are song sections for which notes are repeated in rapid succession; hence, trill rate is the degree to which birds can emit these song sections. Frequency bandwidth is the difference between the highest and lowest discernable frequencies within a song. Among these 34 species, the maximal values of frequency bandwidth decreased with increasing trill rates (fig. 7.4). Most notable was a

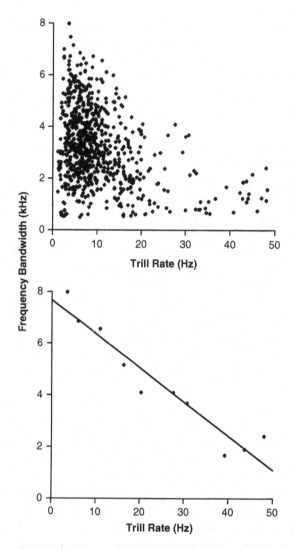

Fig. 7.4 *Upper panel*, Trade-off between trill rate and frequency bandwidth for songbirds. One can observe a notable "triangular" distribution in which maximum values of frequency bandwidth decrease with increasing trill rates. *Lower panel*, Pooled data showing the same pattern, plotted as a linear regression. Taken from Podos (1997) with permission from Wiley-Blackwell press.

"triangle-shaped" pattern between trill rate and frequency. Trills with low trill rates exhibited a wide variance in frequencies used, whereas trills with high trill rates exhibited very narrow frequency bandwidths. This highly bounded pattern suggests that performance constraints have limited the evolutionary diversity of trills and thus bird song.

Mechanical constraints are often exhibited in feeding structures because of the close match between their morphology and the physical shape or the nature of the prey. This is especially true when predators attack and consume prey that possess sophisticated physical defenses. In one species of carabid beetle, *Damaster blaptoides*, there are two primary feeding phenotypes; one type is an elongated form with a small head, whereas the other is stout and has a large head. In experiments with snails, the small-headed morph fed with high efficiency on snails with thick shells and large apertures by using its head to penetrate the snail shell (Konuma and Chiba 2007). An alternative approach was shown by the large-headed beetle, which maximized its feeding performance on snails with thin shells and small apertures by crushing the snails. In other words, there is a trade-off with different feeding strategies; small-headed morphs are effective at penetrating shells, whereas large-headed morphs rely on crushing shells (Konuma and Chiba 2007). This alternative from of performance maximization fits into the general concept of a trade-off between force and the ability to fit into small space. During feeding, the production of high forces usually requires large mouthparts that cannot fit into tight spaces, whereas an agile and small head that can effectively fit into tight spaces usually cannot produce high forces.

The challenge of producing large amounts of force and moving rapidly is one that has the potential to influence many kinds of tasks, but is especially apparent during feeding, as there may be competing pressures on the jaw to be either fast for consuming small rapidly moving prey or forceful for consuming hard prey. In general, wide and deep jaws generate powerful bites for two reasons. First, wider and deeper jaws provide a greater mechanical advantage for producing higher forces than narrower and shallower jaws do. Second, larger jaws can also provide more room for muscle than smaller jaws can and thus results in greater levels of force than can be produced by smaller jaws. However, this increased muscle mass may also mean that jaws cannot be moved as quickly. Thus, wide and deep jaws are particularly good for producing large amounts of force slowly. By comparison, elongated jaws are less able to produce powerful bites but are better suited for rapid movements than short jaws are. These trends have been supported by studies both within and among various animal species, such as in lizards, birds, bats, and crabs. The ecological importance of this trade-off is noticeable for fiddler crabs, *Uca pugnax*, which use their large claws both for feeding and for fighting rivals. Longer claws can be closed with a higher velocity but also have a lower mechanical advantage, that is, they lack the proper physical arrangement to produce high levels of force (Levinton and Allen 2005). In other words, individual crabs cannot be both quick and strong with their claws. This finding suggests that there may be alternative methods for fighting (e.g., quickness vs. strength) within these crabs and also indicates that, as crabs become larger, they gain a crucial advantage in being quicker with their claws, a feature which may be important during fights.

Trade-offs can potentially be ameliorated if a variety of structures that affect performance can be modified independently. Because their swimming ability is influenced by their body shape, physiology, and the shape and size of their many fins, fish body shape is an interesting case in point. Early views of fishes focused on specialization for different functions, such as accelerating (e.g., pike), cruising, or even being a generalist (e.g., trout). Based on a priori biomechanical principles, morphological traits that enable high performance for acceleration should differ greatly from those for cruising. The ability to accelerate in fishes is driven primarily through the presence of a deep and flexible body, which increases thrust, and a large tail. Moreover, because of the need for quick and powerful movements, the primary musculature should be dominated by fast-twitch muscle fibers. On the other hand, fish that are cruisers should possess a stiff body composed largely of slow-twitch muscle, along with a lunate tail, which minimizes drag. There is evidence that specialist fish species are superior for some kinds of performance relative to generalists. However, some fish species seem to possess traits that allow them to be low-speed cruisers but yet not sacrifice their ability to rapidly accelerate. For example, knifefish possess many of the traits associated with cruising yet are very good accelerators. It is possible that their acceleration abilities may be due to their unique method of bending their body during escape. Similarly, pectoral fins may propel a fish during low-speed cruising but also enable high levels of maneuverability depending on the position of these fins on the body (Blake 2004). Certain arrangements of these different morphological traits may enable some fish to both cruise efficiently and move quickly when the time comes.

Because animals have usually evolved to perform with their intact bodies, the loss of body parts can pose a problem. Remarkably, some animals will voluntarily lose their body parts, a behavior known as autotomy. This behavior has typically evolved to enable animals to survive encounters with predators. Autotomy has evolved many times in animals, such as in harvestmen (limb loss), damselfly larvae (tail loss), and lizards and salamanders (tail loss), among others. Autotomy offers a natural experiment for investigating the potentially constraining (or facilitating) roles of key morphological structures, such as limbs or tails. For example, a postanal tail is a key defining characteristic of chordates and is the ancestral state among vertebrates. Within vertebrates, tails play critical roles in locomotion, balance and sexual displays, from fishes to primates. Nevertheless, in many amphibians and reptiles, tails can be voluntarily shed as a drastic yet effective means of escaping predators. Tail loss has been documented in approximately 67% of lizard families and is facilitated through several mechanisms, such as fracture planes between or within vertebrae, as well as a host of other modifications to tail tissues, such as blood vessels and musculature. After separation, the muscles of the autotomized tail can continue to contract, and these movements may distract predators, thereby allowing the lizard to escape (Higham and Russell 2010). There is even evidence that lizard tails may move so vigorously that they ultimately evade the predator. There is actually evidence of lizards returning to the site of autotomy to ingest their autotomized tail! Although this might make your stomach turn, it is quite beneficial for a lizard that had invested a lot of energy into storing fat in the tail—it makes for a nice snack!

Interestingly, the role of tail loss on the most obvious aspect of performance (sprint speed) is ambiguous. Whereas some research shows that loss of the tail enhances sprint speed, other studies show the opposite, and some studies show no effect (McElroy and Bergmann 2013). However, the tail does play a critical role in maintaining stability during jumping in lizards. When green anoles, *Anolis carolinensis*, jump with a complete tail, they exhibit the typical phases of jumping, including takeoff, a suspended phase, and a secure landing. However, once lizards have lost a large portion of their tails (>50%), their jumps rapidly become unstable after a normal takeoff, resulting in the lizards "toppling" backward and often landing on their backs (Gillis et al. 2009; fig. 7.5). Don't worry—the lizards recover nicely from these

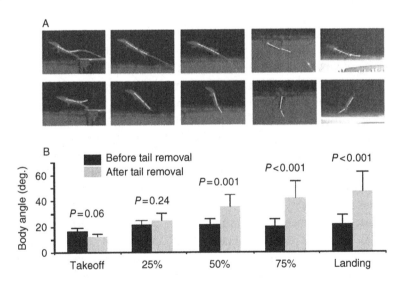

Fig. 7.5 The effect of the tail removal on jumping and stability in the green anole, *Anolis carolinensis*. A, Photos showing the impact of tail removal on the body angle during a jump (from takeoff to landing). B, Histogram of Body angle during jumping, before and after tail removal. Body angle is significantly greater when the tail has been removed, especially later in the jump; this result shows the instability associated with tail autotomy. Graph taken from Gillis et al. (2009) with permission from the Company of Biologists Ltd.

awkward jumps! During the takeoff phase in a lizard with a normal tail, the tail is arched and then slapped down on the jumping surface. Given that tailless lizards tend to rotate backward, this tail-substrate interaction may provide sufficient force to counteract the backward rotation of the body that occurs during jumps, in the same way that a "wheelie wheel" prevents motorcyclists from toppling backward when they perform a wheelie. Thus, the benefits for tail loss must be significant to outweigh these notable costs. While we might pity these poor lizards with their missing tails, it's important to remember that lizards can regenerate their tails, although it may take up to 20 weeks. In at least one species, the leopard gecko, *Eublepharis macularius*, locomotion can be restored after the tail has been fully autotomized and regenerated (Jagnandan et al. 2014).

There exists a broad trade-off within biological musculoskeletal lever systems between the transfer of force and displacement. In other words, those systems that generate force quickly, with high speeds and acceleration, are likely to exhibit decreases in displacement. Suction feeding in fishes provides a nice example of this phenomenon. Suction occurs when a fish rapidly opens its mouth, thereby creating a negative pressure relative to the water outside the mouth. This pressure change causes a rapid influx of water into the mouth, ideally bringing prey as well. In a study of bluegill sunfish, *Lepomis macrochirus*, and largemouth bass, *Micropterus salmoides*, one of us (Tim Higham) and colleagues found that bluegill opened their mouth faster than largemouth bass did, thus generating higher fluid speeds and accelerations (fig. 7.6; Higham et al. 2006). In contrast, largemouth bass, as their name implies, opened their mouths much wider than bluegill did, but fluid speed and acceleration were lower in largemouth bass than in bluegill. This trade-off is likely related to the ecology of these two species. The relatively small mouth of bluegill, coupled with the high fluid speeds and accelerations, are good for picking off small prey items that are not very evasive. In contrast, the large volume ingested by largemouth bass is great for capturing evasive prey while swimming at high speeds: high locomotor speeds during prey capture decrease strike accuracy, so the increased volume comes in handy!

The trade-off between speed and accuracy is pervasive and comes in many flavors. The ability to project tongues at high speeds and accelerations, for example, may come at the cost of poor accuracy. On average, the fastest frog species capture only about 33% of the prey items presented to them, whereas slower species can capture their prey 100% of the time. A similar trend was found for brownsnakes, in that slower strikes were more accurate (Whitaker et al. 2000). This trade-off likely impacts the snake's selection of prestrike display posture, with full displays preceding slower and more accurate strikes. Whether this trade-off occurs generally in animal functional systems has not been examined in

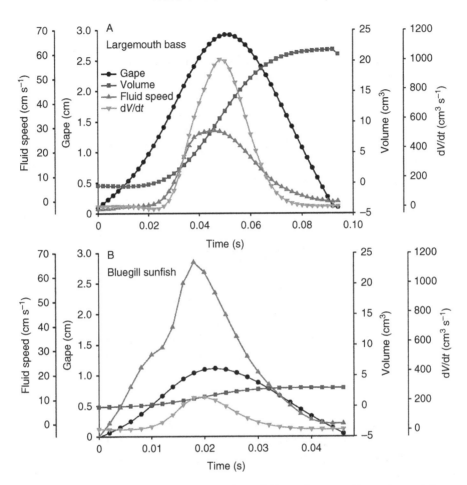

Fig. 7.6 Example sequences of gape, ingested volume, fluid speed generated by suction, and the change in volume with respect to time (d*V*/d*t*) for largemouth bass, *Micropterus salmoides*, and bluegill sunfish, *Lepomis macrochirus*. A, Largemouth bass. B, Bluegill sunfish. A trade-off between maximum fluid speed (higher in bluegill) and ingested volume (higher in largemouth bass) is evident. Also note that maximum mouth diameter (gape) is much greater for largemouth bass than for bluegill. Graph taken from Higham et al. (2006) with permission from the Company of Biologists Ltd.

detail, but it is likely that the ability to exert exacting control over a structure moving at very high rates is challenging, as any baseball pitcher knows.

7.5 Innovations, constraints, and trade-offs

An innovation is a trait (morphological, physiological, or behavioral) that permits animals to explore and occupy new ecological niches and often defines major groups of animals. The defining characteristic of an innovation is that it leads to diversification (increased number of species, greater morphological diversity, or both) within a group. Some examples among vertebrates are limbs, grasping appendages, prehensile tails, and adhesive structures. Although these innovations are extremely beneficial to the animals, they naturally are accompanied by constraints. Let's use a hypothetical example to illustrate this point. Imagine that a group of terrestrial animals exhibited a trait that permitted swimming. This trait would give them access to many new prey items and provide shelter from terrestrial predators (although it might expose them to aquatic ones). However, swimming imposes several hydrodynamic constraints, which might cause evolution in body shape (e.g., streamlining the body to reduce drag), among other traits. In short, novel environments can also impose constraints on evolution. Recent work on gecko adhesion highlights this point. In Namibia, there is a diverse group of geckos from the *Pachydactylus* radiation. This group includes fast diurnal geckos, slow nocturnal geckos, web-footed geckos, burrowing geckos, and rock climbers. An interesting aspect of this group is that some species have secondarily lost or simplified their adhesive system and therefore lost an innovation. Why would a group of species lose such a seemingly valuable innovation? The answer may relate to the nature of trade-offs imposed by toepads.

The adhesive system of geckos involves millions of microscopic hairs on the underside of the toes, as well as underlying anatomical features. This adhesive system harnesses van der Waals forces, which are also commonly used by everyday-life objects, such as tape, although geckos harness these forces in a different way (Autumn and Peattie 2002). Although van der Waals bonds are individually weak, the sum of millions of these weak bonds collectively results in an incredibly strong bond! Unlike other lizards, such as anoles, geckos possess an active adhesive system and must utilize hyperextension (curling up of the toe tips) to disengage the adhesive system so that the foot can be lifted. When the foot reinitiates contact on the next footfall, the tips of the toes are uncurled and placed on the surface on which the animal is moving. This is a very effective mechanism for providing adhesion during climbing, but it also slows the gecko down since it takes time to hyperextend and uncurl the tips of the toes (Russell and Higham 2009; see fig. 7.7). Thus, there is a trade-off between adhesion and running speed. This brings us back to the basis for losing the innovation in the geckos in Namibia. Although most of the geckos from the *Pachydactylus* radiation are climbers, some have moved into terrestrial areas, such as sand dunes, hard-packed sand, or sheet rock. Such habitats obviate the need for climbing and therefore the need for an adhesive system. If adhesion slows geckos down, then the innovation may actually be counterproductive in a terrestrial habitat. In fact, recent work has shown that, once the adhesive system is lost in some of these geckos, the rates of morphological and kinematic evolution actually increase (Higham et al. 2015). Thus,

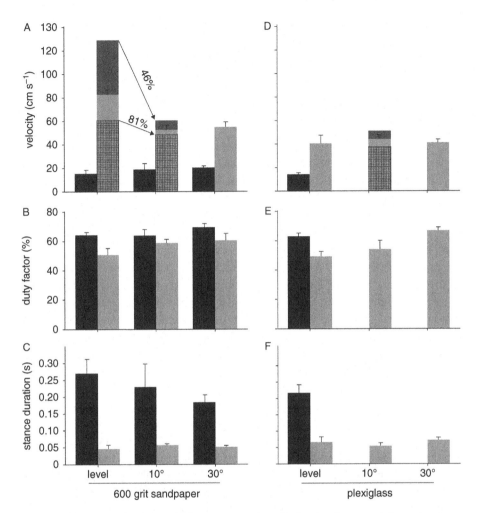

Fig. 7.7 The impacts of incline and adhesion on running velocity, duty factor, and stance duration in *Eublepharis macularius* and *Tarentola mauritanica*; the former species (indicated by black bars) does not have an adhesive apparatus, whereas the latter species (indicated by gray bars) does. Individuals were measured running on a level surface, a 10° incline, or a 30° incline. *A–C*, Individuals running on 600-grit sandpaper. *D–E*, Individuals running on plexiglass. In *A*, dark grey bars represent the mean velocities for *T. mauritanica* individuals that used their adhesive apparatus on the 10° incline; such individuals exhibited a decrease in mean velocity to 46% of the mean velocity achieved on the level surface. Cross-hatched bars represent the mean velocities for *T. mauritanica* individuals that did not use their adhesive apparatus on the 10° incline; such individuals exhibited a mean velocity that was 81% that achieved on the level surface. These results suggest that, in this species, there was a trade-off between adhesion and running speed. Graph taken from Russell and Higham (2009) with permission from the Royal Society of London.

as diversification among these geckos is increasing, losing an innovation may actually be an innovation itself!

7.6 Ecological and reproductive constraints on performance

Another form of ecological constraint arises from the necessary function of reproduction. Reproduction is a vital task; yet, for many animals, reproducing places the sex that bears the offspring (usually the female) at considerable risk. The degree of risk varies among different species according to parental investment. In some species, eggs can take up an enormous amount of physical space inside the female, and the process of raising the eggs to maturity can be physically exhausting. In some cases, the sheer mass of the eggs or offspring directly causes a decrement in locomotor performance. For example, in the common iguana, *Iguana iguana*, females can gain up to 60% of their body weight in eggs, and this increased load dramatically diminishes their maximum sprint speed. Manipulative studies in zebra finches have shown that reproduction can also cause a diminishment of muscle mass that persists even after the extra mass of eggs is relieved, resulting in a decline in takeoff speed. Therefore, the negative impacts of reproduction on performance may linger after the burden of reproduction has passed. Further, there may also be trade-offs between how fecund a female is and how much she declines in locomotor performance, both when possessing eggs and after egg laying. The effect of reproduction in relation to performance has also been examined in studies of natural selection. Like many lizards, female side-blotch lizards can produce variable numbers of eggs, and as noted above, large egg loads appear to diminish locomotor performance in lizards. This may explain why natural selection is directional in side-blotch lizard males, favoring only the fastest individuals, whereas it is stabilizing in females, favoring average performers. Stabilizing selection is somewhat rare in nature, but the large egg burdens may prevent females from achieving very high speeds; thus, females with relatively few eggs would not be favored by natural selection. In ascidian larvae, the evolution of reproductive effort seems closely tied to the evolution of clonality and locomotor performance. It appears that the ability to be clonal has evolved several times, and clonal larvae seem to possess a relatively large trunk and a relatively short tail. A shorter tail seems to make swimming more energetically efficient. As these animals become clonal, they also seem to have evolved larger body size, although size appears to have little impact on their locomotion.

Constraints can also arise from reproduction, by forcing animals to evolve behaviors or morphological traits that are necessary for mating. For example, sexual selection can induce the evolution of mating structures that can dramatically hinder performance. In some cases, animals have evolved behaviors that mitigate this constraint. The cobweb spider *Tidarren sisyphoides* is a remarkable example of sexual size dimorphism. Spiders are renowned for females exceeding males in size but, even among spiders, this species is extreme! Females can be up to 100 times larger than males, presenting an extreme challenge for mating. This challenge comes in two forms; first, the much smaller male must mate with the female, and therefore males must have genitalia that are sufficiently large to couple with the female and transfer sperm. Second, upon their penultimate (next to last) molt, the much smaller male must find receptive females before other males arrive, to ensure that the female will carry his offspring. This second task is confounded by the extremely small size of the male. Consider that a male *Tidarren* spider moving even 1 m is equivalent, in terms of body lengths, to a 3 m alligator moving about 5 km. Hence, for such excessively small male spiders, effective locomotion is crucial to their mating success. However, the locomotion of male *Tidarren* spiders is greatly impaired by two extraordinarily large pedipalps, which are modified limbs that are used as reproductive organs (fig. 7.8). These excessively large pedipalps can weigh up to 10% of the body mass of the spider and can overlap one another.

Surprisingly, almost immediately after their penultimate molt, these spiders attach one of their two pedipalps to a silk thread and twist in circles until the reproductive organ is removed. Why would these spiders exhibit such an extreme behavior? By measuring both maximum speed and maximum endurance of male spiders before and after pedipalp removal, the value of this behavior

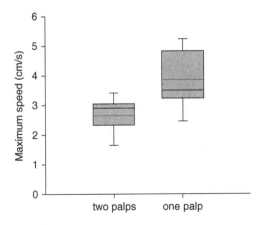

Fig. 7.8 Graph showing the maximum speed of male spiders, *Tidarren sisyphoides*, before and after pedipalp removal. Speed increased by 44% after removal, an increase that was statistically significant. Graph taken from Ramos et al. (2004) with permission from the National Academy of Sciences.

becomes clearer. Maximum speed is enhanced by about 44% by pedipalp removal, and maximum endurance nearly triples (Ramos et al. 2004). The most likely explanation for this bizarre behavior is that, when both of the overly large pedipalps are present, they drag on the surface, thus impairing movement. By removing one pedipalp, the other can be lifted off the ground, thus enabling the spider to move faster and for longer periods of time. Why have these spiders not simply evolved one pedipalp? While the answer is not currently known, one possibility is that coupled appendages, such as pedipalps, are likely more difficult to alter evolutionarily compared to the evolution of a novel behavior. Thus, this case is an example of how the pressure of reproduction has resulted in the evolution of a novel performance-enhancing behavior.

7.7 Overcoming trade-offs: The role of behavior

So far, we have emphasized some of the more inexorable mechanical trade-offs imposed on organisms because of the relatively inflexible nature of anatomy. However, the role of behavior is less of a focus in the discussion of trade-offs. Animals have great flexibility to move their bodies in a variety of ways, especially if they have multiple body parts that influence the performance capacity in question. Performance abilities are not always created in a stepwise fashion, with each morphological part adding in a linear fashion. Moreover, the view that the "sum is greater than the whole of the parts" is an apt one for performance traits. Studies by Sharon Emerson and colleagues on flying frogs have nicely demonstrated this point. Some species of tropical frogs have evolved elaborate webbed feet that enable the frogs to glide among trees. These frogs have several other morphological specializations that appear to enhance gliding, such as lateral skin flaps and a relatively lean build. Gliding performance is typically defined as the total horizontal distance a frog could glide for a given vertical height, but there are other metrics which may be important to a glider, such as how much time they spend aloft, how fast they move, and so forth. Because both the shape of the different morphological elements and how these elements are used (i.e., posture) are likely to affect gliding performance, the researchers examined the interactive roles of behavior and morphology on gliding in these animals using both live and fake (model) frogs (Emerson and Koehl 1990). They found that one could not simply deconstruct the different components of a frog and add up their individual effects to understand the gliding performance of the whole organism. Instead, flying frogs exhibited emergent behaviors that allowed them to alter the relationship between morphology and gliding performance: the typical flying posture of flying frogs appears to decrease the total gliding distance but improve maneuverability. Thus, for these animals, this aspect may play the most important role in determining gliding performance.

References

Autumn K, Peattie AM. 2002. Mechanisms of adhesion in geckos. Integrative and Comparative Biology 42:1081–1090.

Biewener AA. 1990. Biomechanics of mammalian terrestrial locomotion. Science 250:1097–1103.

Blake RW. 2004. Fish functional design and swimming performance. Journal of Fish Biology 65:1193–1222.

Emerson SB, Koehl MAR. 1990. The interaction of behavioral and morphological change in the evolution of a novel locomotor type: "Flying" frogs. Evolution 44:1931–1946.

Garland T Jr. 1988. Genetic basis of activity metabolism. I. Inheritance of speed, stamina, and antipredator displays in the garter snake *Thamnophis sirtalis*. Evolution 42:335–350.

Garland T Jr. 1994. Quantitative genetics of locomotor behavior and physiology in a garter snake. In Boake CRB, ed. Quantitative Genetic Studies of Behavioral Evolution. University of Chicago Press, Chicago, IL, pp. 251–276.

Gillis GB, Bonvini LA, Irschick DJ. 2009. Losing stability: Tail loss and jumping in the arboreal lizard *Anolis carolinensis*. Journal of Experimental Biology 212:604–609.

Gould SJ, Lewontin RC. 1979. The spandrels of San Marco and the Panglossian paradigm: A critique of the adaptationist programme. Proceedings of the Royal Society of London B 205:581–598.

Grant PR, Grant RB. 2014. 40 Years of Evolution: Darwin's Finches on Daphne Major Island. Princeton University Press, Princeton, NJ.

Greer-Walker M, Pull GA. 1975. A survey of red and white muscle in marine fish. Journal of Fish Biology 7:295–300.

Herrel A, Podos J, Vanhooydonck B, Hendry AP. 2009. Force-velocity trade-off in Darwin's finch jaw function: A biomechanical basis for ecological speciation? Functional Ecology 23:119–125.

Higham TE, Birn-Jeffery AV, Collins CE, Hulsey CD, Russell AP. 2015. Adaptive simplification and the evolution of gecko locomotion: Morphological and biomechanical consequences of losing adhesion. Proceedings of the National Academy of Sciences 112:809–814.

Higham TE, Day SW, Wainwright PC. 2006. Multidimensional analysis of suction feeding performance in fishes: Fluid speed, acceleration, strike accuracy and the ingested volume of water. Journal of Experimental Biology 209:2713–2725.

Higham TE, Russell AP. 2010. Flip, flop and fly: Modulated motor control and highly variable movement patterns of autotomized gecko tails. Biology Letters 6:70–73.

Hudson PE, Corr SA, Payne-Davis RC, Clancy SN, Lane E, Wilson AM. 2011. Functional anatomy of the cheetah (*Acinonyx jubatus*) hindlimb. Journal of Anatomy 218:363–374.

Jagnandan K, Russell AP, Higham TE. 2014. Tail autotomy and subsequent regeneration alter the mechanics of locomotion in lizards. Journal of Experimental Biology 271:3891–3897.

Konuma J, Chiba S. 2007. Trade-offs between force and fit: Extreme morphologies associated with feeding behavior in carabid beetles. American Naturalist 170:90–100.

Lailvaux SP, Husak JF. 2014. The life-history of whole-organism performance. Quarterly Review of Biology 89:285–318.

Levinton JS, Allen BJ. 2005. The paradox of the weakening combatant: Trade-off between closing force and gripping speed in a sexually selected combat structure. Functional Ecology 19:159–165.

McElroy E, Bergmann PJ. 2013. Tail autotomy, tail size and locomotor performance in lizards. Physiological and Biochemical Zoology 86:669–679.

Oufiero CE, Walsh MR, Reznick DN, Garland T Jr. 2011. Swimming performance trade-offs across a gradient in community composition in Trinidadian killifish (*Rivulus hartii*). Ecology 92:170–179.

Podos J. 1997. A performance constraint on the evolution of trilled vocalizations in a songbird family (Passeriformes: Emberizidae). Evolution 51:537–551.

Ramos M, Irschick DJ, Christenson TE. 2004. Overcoming an evolutionary conflict: Removal of a reproductive organ greatly increases locomotor performance. Proceedings of the National Academy of Sciences 101:4883–4887.

Reidy SP, Kerr SR, Nelson JA. 2000. Aerobic and anaerobic swimming performance of individual Atlantic cod. Journal of Experimental Biology 203:347–357.

Rome LC, Funke RP, Alexander R McN, Lutz G, Aldridge H, Scott F. 1988. Why animals have different muscle fibre types. Nature 335:824–827.

Russell AP, Higham TE. 2009. A new angle on clinging in geckos: Incline, not substrate, triggers the deployment of the adhesive system. Proceedings of the Royal Society B 276:3705–3709.

Vanhooydonck B, Van Damme R, Aerts P. 2001. Speed and stamina trade-off in lacertid lizards. Evolution 55:1040–1048.

Whitaker PB, Ellis K, Shine R. 2000. The defensive strike of the eastern brown-snake, *Pseudonaja textilis* (Elapidae). Functional Ecology 14:25–31.

Wilson RS, James RS, Van Damme R. 2002. Trade-offs between speed and endurance in the frog *Xenopus laevis*: A multi-level approach. Journal of Experimental Biology 205:1145–1152.

8 | Sexual selection and performance

8.1 What is sexual selection?

Many people have probably had the experience of walking through a park or zoo and encountering a beautiful male peacock adorned with long, colorful, iridescent, and extravagant feathers that drooped on the ground as the bird meandered about. Perhaps you have even asked whether this is a bit much? One of the authors of this book (Irschick) has studied the theory of sexual selection for many years and was familiar with examples of such ornaments; but seeing a male peacock elicited this kind of reaction, even though he had seen peacocks before. And therein lies the crux of the issue that has bedeviled scientists for many years. Why do some animals possess such colorful, extravagant, and apparently useless or even detrimental physical structures? The peacock tail is one example of a broader class of morphological structures or patterns called sexually selected traits. As noted above, sexually selected traits are frequently present or exaggerated in only one sex (frequently males), and they are distinct from other morphological traits in that they are primarily important during two processes important for reproduction: male competition, and female choice (Maynard-Smith and Harper 2003). Male competition refers to the process by which males compete with one another for access to resources, territories, or mates (or all of these) and is often manifested by males vigorously fighting or signaling to acquire a certain spot of earth or to acquire a vital resource. During such fights, males often use their sexually selected traits to signal various measures of quality, such as strength, size, vigor, or athletic ability. The sexual structures are also sometimes used directly in the fighting process; for example, antlers in ungulates are used both to signal male fighting ability and to actually fight (Emlen 2014).

Animal Athletes. Duncan J. Irschick and Timothy E. Higham. © Duncan J. Irschick and Timothy E. Higham 2016. Published in 2016 by Oxford University Press.

Why should we care about sexually selected traits, and what is their relevance for performance traits and animal function? Let's address the first part of this question. Sexually selected traits are interesting for many reasons, but the foremost is that they provide an opportunity to study trade-offs, as in many species such traits appear to impair other vital functions, such as predator escape. In other words, how can we explain the presence of structures that in many cases seem to oppose the struggle for survival, not aid it? The answer resides in trade-offs between the desire to mate, and thereby leave offspring that will pass on your genes, and the ability to simply survive. Sexual structures, though seemingly superfluous, enhance the ability to compete with the opposite sex for access to mates or attract members of the opposite sex. Much of the benefit of sexually selected traits can be broken into separate functions for male competition and female choice. Sexually selected traits are also important when females are attempting to choose males, although in some species, this role is reversed. Males will sometimes display the size, shape, and color of their sexually selected traits to demonstrate to females the same qualities that they display to rival males (Searcy and Nowicki 2005). Sometimes, male competition and female choice may be intermingled, such as when females choose the winners of male fights, or when females choose among males that simultaneously display in groups (leks), much like a dancing competition with a skeptical female audience. In sum, sexually selected traits are in many cases transmitters of information about the underlying quality of the individual; however, what "quality" means is often vague and poorly defined, and in this regard performance traits offer some striking advantages.

8.2 A functional approach to sexual selection

Until fairly recently, the study of sexual selection has taken place largely independently of functional studies, but recent efforts to integrate these fields have reaped some benefits (Irschick et al. 2015). Let's consider two male deer fighting one another to gain access to a group of females. There are many factors that will dictate which male will win, but some obvious ones include male body size, male endurance capacity, male strength, and aggression. It's important to note that practically all of these factors, among others, can be quantified as viable performance traits, in much the same manner as burst swim speed in a mako shark, or jumping distance in a lemur. By measuring performance traits in each of these male deer and investigating how they affect which male deer wins, we can apply a more rigorous and quantitative methodology for defining male "quality," especially if we can link "quality" to reproductive success. In other words, we can address the simple question of whether high-performance males enjoy a reproductive or competitive advantage, either because they more often win fights or because they are preferred by females (or both). This functional approach toward sexual selection is intuitive, as many of the concepts in sexual selection theory are rooted in physiological and functional concepts.

Let's take a brief tour of some of the main ideas that are debated in the field of sexual selection theory. One idea is that "high-quality" males will enjoy a significant advantage both when competing against rival males and when attracting females. This idea has been formulated to mean that males that possess "good genes" are those of the highest quality and that females prefer to mate with such males because they want their own offspring to possess these genes (Hunt et al. 2004). In practice, researchers are rarely able to directly measure genetic "quality," and instead focus on more measurable measures of quality that are likely to be correlated with genetic quality. This idea is based on the premise that resources are rarely sufficient to support an unlimited number of males in a population, especially as females are typically choosy about with whom they mate. However, high-quality males may not be superior in every aspect, and sexual selection theory has modeled some leeway for trade-offs with other traits. For example, males that invest heavily in structures or behaviors that enhance mate acquisition may compromise physiological or functional processes that favor survival and thus have a reduced life span. A second idea is that sexual signals transmit information from the signaler to the receiver, such as a female, a rival male, or a predator. This idea is often termed "honest signaling," which means that males that possess signals that are unusually large, colorful, extravagant, or some combination thereof are of particularly high quality (Zahavi and Avishag 1997). In some cases, signals can be dishonest, meaning that a male is bluffing about its intrinsic qualities, but examples of such signals are rare. These ideas allow us to explore some fascinating questions in relation to animal performance and function. Are males with high performance capacities more often successful when competing against rival males? Are whole-organism performance traits good metrics of male quality? Finally, is there an innate female preference for high-performing males, or counterintuitively, do females prefer wimpy performers?

8.3 Male competition and performance

The value of using performance traits to understand the outcomes of male fights seems intuitive. When males fight for access to females or territories, it is fairly easy to determine the "winner" and compare its morphology and performance traits to that for the "loser." Indeed, in small invertebrates and vertebrates, one can easily stage male fights in the laboratory and measure the relevant variables prior to the fight. In smaller animals such as fish, insects, or lizards, one can stage fights by placing two or more males in an arena and determining the winner in the subsequent battle, although one must always keep in mind the staged nature of these interactions. Larger animals such as ungulates are harder to manipulate in this manner, but one can gain information about such males as part of a larger demographic study and thus compare various traits between winners and losers.

Some taxonomic groups have received a great deal of attention because they are easier to manipulate in artificial settings. Lizards are well studied in terms of male competition, because many species are territorial and will actively fight rival males. Scientists continue to debate the exact nature of how animals "assess" one another before embarking on a full-fledged fight that is likely to hurt one or both combatants (Andersson 1994; Lailvaux and Irschick 2006a). For most animals, male competition is frequently less violent (at least initially) than is often depicted on TV or cable shows and usually begins with a ritualized mutual assessment in which males display at one another from a distance; this initial assessment is followed by increasingly more aggressive behaviors, which ultimately conclude with biting, striking, or grappling, during which one or both competitors can be injured. This process, known as the sequential assessment game, can be abbreviated if one of the males decides that the rival is too formidable and retreats. Lizards follow this general model closely. Rivals typically begin fights with several kinds of displays, such as head-bobbing, pushups, and the extension of an enlarged flap of skin (termed a dewlap), which often reveals striking colors. If these displays are unsuccessful, males often resort to biting, chasing, and grappling. This implies that the ability to bite hard and run fast should be strong predictors of fight success in lizards, and the available data back this assumption. In a variety of lizard species, male lizards that are hard biters more frequently win fights and, in some cases, have enhanced fitness. For example, in collared lizards, the home ranges of male lizards with high bite forces tend to overlap with an increased number of female home ranges, thus providing greater levels of reproductive opportunity (Lappin and Husak 2005). Additionally, maximum running speed and locomotor endurance both appear to play a role. In the lizards *Sceloporus occidentalis* and *Anolis cristatellus*, males with high sprint speeds (*S. occidentalis*) or high endurance capacities (*A. cristatellus*) are more likely to win dyadic male-male contests than are low-performance males. Further, in the lizard *Urosaurus ornatus*, both sprint speed and endurance predict male dominance.

The role of performance traits and sexual structures during male fights also varies among social and age classes. In most animals that fight, how big a competitor is means a lot. Unlike organized fighting in humans, most animals are not strict about "weight limits," although it is unusual for extremely small males to fight much larger ones. However, it is not uncommon to observe males with quite different morphological traits competing against one another. Males often form distinctive groups, known as morphs, which can differ morphologically and behaviorally. These morphs will frequently compete with one another for access to females, using tactics based on differing performance capacities. In some cases, these morphs are of approximately similar ages and remain fixed for life. For example, in the side-blotch lizard, *Uta stansburiana*, there exist three alternative "morphs" that coexist side by side (Sinervo and Lively 1996). Each morph employs a different mating tactic. The territorial orange-throated males possess higher endurance capacities than either the yellow-throated sneaker or blue-throated guarder males. Their names describe their various mating tactics. One morph specializes in defending territories, another in sneaking matings, and the final morph in guarding mates. It is also possible that higher endurance capacities provide a contest advantage for territorial orange-throated males, as male fights can be exhausting (Sinervo et al. 2000). Within pupfish, *Cyprinodon pecosensis*, territorial males also exhibit higher swimming endurance than nonterritorial males do (Kodric-Brown and Nicoletto 1993).

Agility can vary with body size, and therefore males of different sizes might also vary in the degree to which agility is used as a fighting strategy. Hornless sneaker males of the dung beetle *Onthophagus taurus* morph often maneuver in tunnels when competing for females against more formidable horned guarder males (Moczek and Emlen 2000). This contrast between reliance on force for more formidable morphs and on quickness for less formidable morphs is also seen in green anole lizards, *Anolis carolinensis*, for which bite force dictates fight success in hard-biting large males, whereas jumping ability dictates fight success in smaller and quicker individuals (Lailvaux et al. 2004). The overarching role of bite force for larger green anole morphs makes sense because animals bite harder as they mature; thus, its use as a tactic becomes more potent as animals become larger.

8.4 Are sexual signals honest signals of male performance?

As noted above, an important component of male fighting is the use of sexual signals that may relay information about intrinsic male quality. In fact, such signals are important more broadly in the realm of male territorial defense, not just during fighting. Sexual signals come in many varieties, but one crucial distinction is between armaments, which are structures used directly in fights, and ornaments, structures whose sole function is to signify intrinsic information about the individual. Examples of the former include antlers in elk, and horns in beetles, each of which is used directly during fights. Some common kinds of ornaments include color patches or modifications of the skin or body that make the animal appear larger or more formidable. A sexual signal need not be a unique structure but rather can be an enlarged version of a more generalized structure, such as an enlarged head in a lizard that bites rivals.

One version of the handicap hypothesis posits that some sexual signals or sexual behaviors are so exhausting to use or to possess that they can only be exhibited by high-quality males. There is some support for this view. Many animals communicate with rival males and females through the use of visual displays, many of which seem designed to showcase athletic abilities. In the lizard *Uta stansburiana*, individuals that are forced to undergo intense exercise are generally unable to perform "pushup" displays, which are regularly used for territorial advertisement (Brandt 2003). This result suggests that pushup displays are energetically exhausting. There is also evidence that the ability to display vigorously to predators is correlated with enhanced endurance capacities in other lizard species, such as in the crested anole, *Anolis cristatellus* (Leal 1999). Thus, males that display frequently may be confident in their intrinsic abilities to elude or fight a predator if things take a turn for the worse. While it might seem a suicidal strategy for us, it is not unusual for prey to openly confront predators, which amounts to the principle of "I see you, don't bother chasing me!"—also known as the pursuit-deterrent hypothesis. However, this finding with crested anole lizards adds a twist to this idea, as it suggests that the behavior of confronting predators may be more likely to be used by high-performance individuals.

However, honest signaling can also be reflected in the relative size, color/brightness, and shape of the sexual signal. In several highly territorial *Anolis* lizard species, the relative size of the dewlap is positively correlated with bite force (Vanhooydonck et al. 2005). Territorial anoles often bite one another during male fights and, in laboratory trials, those males that are especially hard biters more often win fights (fig. 8.1). Such fights can be dangerous for the combatants, particularly for those males that cannot inflict as much damage as their rivals. The dewlap may thus act as a long-distance signal that displays male fighting prowess to rivals (fig. 8.2). However, in less-territorial anole species, there is no evidence that dewlap size honestly signals bite force, nor is there evidence that possessing high bite forces is instrumental to fight success (Lailvaux and Irschick 2007). Therefore, the role of performance traits for dictating male fight success varies among species in accordance with social context.

Within invertebrates, there is also evidence that male armaments may signal some performance traits that are important during fights. Dung beetles, *Euoniticellus intermedius*, fight by grappling with one another, aided by their horns, and those males with relatively large horns also have higher endurance capacities; although there is no direct evidence that endurance allows males to win fights, this linkage is suggestive. The relative size of the horns in these

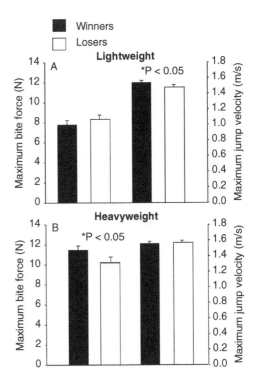

Fig. 8.1 *Left*, Image of a green anole *Anolis carolinensis*. *Right*, Histograms showing the role of bite force and jumping ability for dictating fight success during staged encounters in green anoles. *A*, Role of bite force and jumping ability in fight success for small (lightweight) green anoles. *B*, Role of bite force and jumping ability in fight success for large (heavyweight) green anoles. For large (heavyweight) male green anoles high bite forces provide an advantage during fights, whereas for smaller (lightweight) male green anoles quickness (i.e., jumping ability) provides an advantage for fighting. Graph taken from Lailvaux et al. (2004) with permission from Royal Society Publishing. Image of *A. carolinensis* is by Duncan J. Irschick.

beetles is also a predictor of how much force is required to pull a beetle out of a hole (Lailvaux et al. 2005). When one considers the manner of fights in dung beetles, both of these performance traits, endurance and the ability to resist pulling, offer some advantages. Dung beetles undergo prolonged fights during which they attempt to pull one another out of small tunnels beneath the dung "patties," and it is possible that males can assess one another by examining the size of their horns.

On the flip side, just as sexual structures can be "honest" signals of animal performance traits, males can also "cheat" by relaying false information about their status, otherwise known as dishonest signaling. The most common method of dishonest signaling is when males possess a signal or structure that normally signifies a high level of quality but in fact are wimps. Theoretical

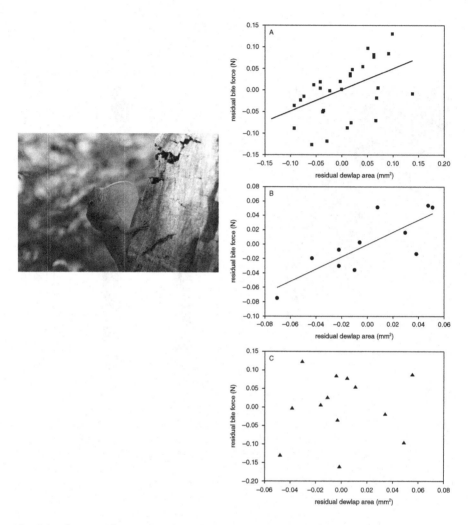

Fig. 8.2 *Left*, Image of an *Anolis cybotes* displaying its dewlap. *Right (A–C)*, Relationship between residual dewlap area and residual bite force for three anole lizard species varying in social behavior: A, For *A. lineatopus*. B, For *A. grahami*. C, For *A. valencienni*. For the highly territorial *A. lineatopus* and the territorial *A. grahami*, there is a significant and positive relationship between the two variables; these results indicate that, in these species, dewlap area is an honest signal of fighting ability. By contrast, in the nonterritorial *A. valencienni*, there is no relationship between these variables; this result indicates that dewlap area is not an honest signal in this species. Graph taken from Vanhooydonck et al. (2005) with permission from Wiley-Blackwell Press. The image of *A. cybotes* was taken by Duncan J. Irschick.

models predict that cases of dishonest signaling should be rare for several reasons. First, the possession or use of a signal may be energetically expensive, and only high-quality males should be capable of such displays. A second reason is that, if rival males choose to challenge the cheater, then the ensuing fight could be dangerous for the cheater. Despite these predictions, cheaters do exist in some instances. In crayfish, males attack one another using their claws and, in general, the larger and wider the claw, the more potent it is as a weapon (Wilson and Angiletta 2015). Crayfish also use their claws as signals, as they wave them in front of other males. A biomechanical analysis indicates that, for "closing" structures (claws or jaws), there are two primary factors: closing speed and closing force. For crayfish, the former is most important. However, while the general relationship between claw size and claw function is upheld for most crayfish, some individuals cheat because they possess enlarged claws that are gracile, wimpy at pinching, and therefore ineffective during fights, or because they might display regenerated claws with little muscle. What happens to these bluffers that do get in real fights is poorly understood, but this is a compelling example of cheating (fig. 8.3).

8.5 Sexual selection imposing costs on performance

Up to now, the discussion has centered on the premise that only high-quality males have especially enlarged, elaborate, or colorful signals and that these signals are usually honest metrics of quality. However, this assumption must be counterbalanced against potential costs associated with the possession of colorful or extravagant sexual traits. It seems obvious that some kinds of sexual structures would seem to impose an obvious cost. Males of many bird species, especially those in tropical environments, possess colorful and elongated tail feathers that are desirable to female birds (i.e., the longer the better). However, elongated tail feathers also impose a potential cost for flight. Based on both theoretical models and wind-tunnel data, some birds with especially elongated tail feathers suffer a substantial decrement in flight performance (fig. 8.4; Balmford et al. 1993), although some species may have compensatory mechanisms that eliminate such declines in performance. This observation leads to an obvious question: if the possession of tail feathers is at times

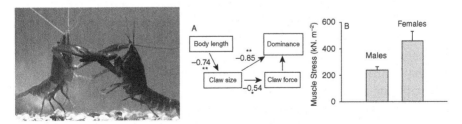

Fig. 8.3 *Left*, Image of crayfish fighting. *Right*, Relationships between morphology and physiology in relation to dominance in crayfish. *A*, Dominance in crayfish is defined primarily by chela size ($P <$ 0.001). *B*, Histogram of chela muscle stress in male and female crayfish; chela muscle stress in males was almost twice that of females. Graph taken from Wilson et al. (2007) with permission from the University of Chicago Press. Photo is from Anthony O'Toole.

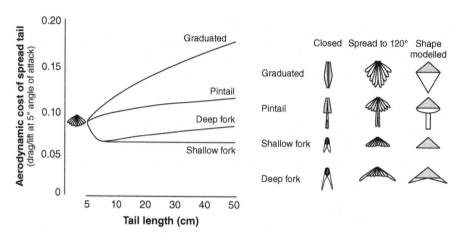

Fig. 8.4 *Left*, The flight costs, in terms of aerodynamics, that are correlated with the elongation of tail feathers, based on modeling data. All models used a simple tail in which all feathers were 5 cm long. All tails were spread to an apical angle of 120°For calculation of lift and drag. *Right*, The tail shapes examined and the corresponding geometric polygons used in the model. See Balmford et al. (1993) for more details. Redrawn from Balmford et al. (1993) with permission from the Nature Publishing Group.

deleterious for flight performance, why has it evolved? The answer lies in a fascinating preference by females for elaborate sexual structures, even if the possession of those structures is harmful to the bearer. An extreme form of this process is termed runaway selection, in which an initial female preference for a novel male trait results in males that possess it being favored by sexual selection. This process of linked female preferences and male traits continues until the costs from possession of the trait clearly outweigh the benefits from enhanced reproductive success. In this manner, some extravagant and down-right ridiculous structures can evolve because they have not yet reached that tipping point.

However, performance costs or trade-offs may also be manifested in less spectacular ways. Even though enlarged horns in dung beetles are honest signals of endurance capacity and pulling strength, they also impede the ability of these beetles to move in the tight confines of the tunnels that these beetles use to move about, thus offering an advantage to the hornless "sneaker" males. Another example is the stalk-eyed fly, the males of which possess elongated eyes that protrude on stalks. These flies have a peaceful yet vigorous method for comparing one another. They line up eye to eye and compare the relative lengths of their eyestalks; those males with the longer stalks are typically the winners. These exaggerated eye stalks seem to increase the moment of inertia for males as compared to females; however, interestingly, males do not ex-hibit any decrement in flight performance, in part because males in large-eyed species also possess larger wings than females do (Husak et al. 2011). These

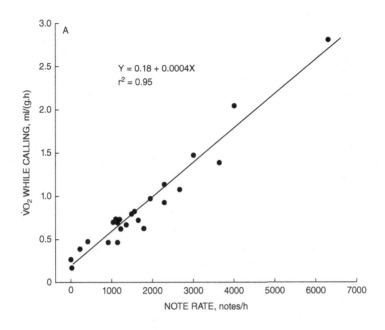

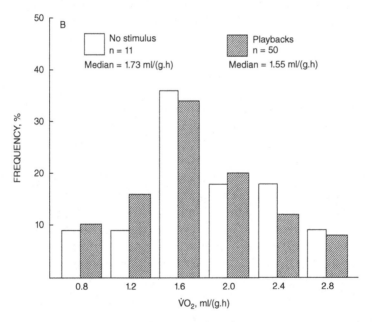

Fig. 8.5 *A*, A plot showing the relationship between note rate and the rate of oxygen consumption for the frog *Hyla microcephala* when the frog is vocalizing. *B*, A frequency distribution of different oxygen consumption rates (VO$_2$) for the frog during vocalization. The open and filled bars indicate vocalizations in response to playback or without a stimulus, respectively. Note that, in *A*, the higher the note rate, the higher the rate of oxygen consumption. Redrawn from Wells and Taigen (1989) with permission from Springer publishing.

compensatory mechanisms appear to be widespread and can ameliorate the burden of sexually selected traits.

The issue of costs is especially relevant for signaling that is energetically exhausting. In frogs and toads, males vocalize to attract females. In general, those males that can vocalize the longest greatly increase their odds of mating. Vocalization in male frogs and toads is driven by muscles which are specialized for aerobic metabolism. Metabolic rates can increase by one- to twofold compared to resting metabolism (fig. 8.5). Because of the expensive yet necessary nature of calling, male frogs that experience prolonged breeding seasons will lose a fair amount of body mass over time. Thus, breeding in frogs and toads presents a trade-off. If you want to mate, you need to expend energy, and those males that possess the energetic reserves, the physiology, or both to do so are more likely to increase their fitness. Moreover, even when a male frog can entice a female to his vicinity, he still faces the energetic cost of mating. In some species, mating is expensive, as males use amplexus, a form of mating in which the male tightly grasps the wet and wiggling female tightly with the forelimbs while simultaneously fending off rival males! The costs of vocalization in relation to mating are not confined to frogs and toads. For example, in tree crickets, males raise their metabolic rates 6–16 times above resting levels when performing stridulations (Prestwich and Walker 1981). Females may use the level of mating display as an index of male quality, if only high-quality males can afford to signal for long periods of time. Wolf spiders, *Hygrolycosa rubrofasciata*, perform a mating display of drumming dry leaves with their abdomens (Kotiaho 2000). These drumming displays are energetically expensive, as maximum metabolic rates can be up to 22 times higher than resting metabolism (fig. 8.6). Indeed, males that are induced to drum frequently rapidly lose body weight. How long a male spider can signal is driven by many factors, but one factor is the amount of food they consume. Well-fed male wolf spiders can drum longer than food-limited males. Much like the example of females frogs and toads preferring those males that sing the longest, female wolf spiders prefer males with high drumming rates. This trait varies substantially among males, and is repeatable, fulfilling two important criteria for this trait to evolve via sexual selection. This example shows the condition-dependent nature of signaling performance and its energetic costs. Those males in favorable ecological contexts are less likely to suffer the energetic consequences of signaling rapidly and often.

8.6 Female choice and performance

The interrelationship between sexual selection and performance traits makes sense in the context of male competition. The role of female choice would seem equally straightforward. Nearly every nature TV show expounds the view that females prefer to mate with the most vigorous and healthy males,

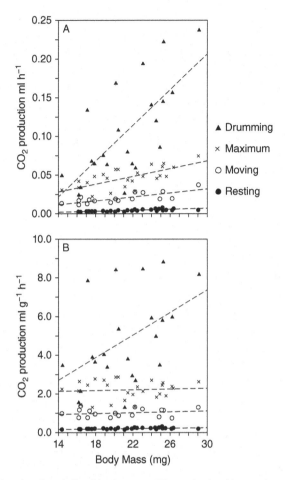

Fig. 8.6 Graphs showing the relationships between CO_2 production (*A*, raw data; *B*, scaled data) and mass for drumming, maximal exertion, moving, and resting in the wolf spider, *Hygrolycosa rubrofasciata*. Taken from Kotiaho et al. (1998) with permission from the Royal Society Publishing Group.

but is this view correct, or is it based on our biased view of how nature should work? Once one digs a little deeper, this expectation is far from obvious, and the available data support this ambiguity, at least so far. For example, males that invest substantially in sexual structures may simply be good-looking and not very vigorous and, in some species, females may prefer to mate with males that can assist in caring for their offspring. While it is intuitively appealing that female animals should prefer to mate with the strongest or most vigorous male, doing so might expose females to bodily harm, parasites, or many other ailments. Thus, in some cases, females might choose to mate with males who are wimpy performers and are less likely to display violent behavior. Unfortunately, there is little data that test these ideas. The reasons for this lack of knowledge are severalfold. While male fights are relatively easy to stage in the laboratory and interpret, female choice experiments are experimentally challenging, as multiple interacting factors, such as body size, age, color, and morphological shape, may influence female choice.

In at least a few cases, there is some evidence that females prefer high-performance males. In some guppy species, females prefer males with enlarged and colorful spots (Kodric-Brown and Nicoletto 2005). The bright colors found in these spots are partially derived from carotenoid pigments that are derived from the food the guppies consume, such as fruit or vegetables. Thus, males that are successful at eating high-quality food types rich with carotenoids are more likely to have large and colorful spots. In the guppy, *Poecilia reticulata*, there is a positive relationship between carotenoid pigment density and male swimming endurance, suggesting a general linkage between male vigor and the morphology of sexual ornaments. Additionally, swimming endurance is correlated with ornament complexity, defined as the number of color spots and the degree of orange coloration (fig. 8.7). Females prefer to mate with brightly colored males; therefore, this interlinking of male performance with ornament morphology raises the question of whether females prefer to mate with high-performance males or whether this preference is a correlated response to female preference for males with large sexual ornaments. A more recent study found that the ability to evade a predator for a given guppy is positively related to the coloration of its father; having a father who is a good swimmer offered no advantages for prey avoidance (Evans et al. 2011). In other words, while there is some evidence for the role of swimming performance for influencing female choice in guppies, it's likely that there are other factors that influence female choice as well. At least one other study shows a more ambiguous result. In choice tests with female green anoles there was no clear preference by females for males based on any number of traits, including bite force, jumping ability, dewlap size, and other morphological traits (Lailvaux

Fig. 8.7 An image of a guppy, *Poecilia reticulata*. Image by Astrid Kodric-Brown.

and Irschick 2006b). Whether this lack of preference was due to female disinterest in these traits, the unnatural aspect of the laboratory design, or the possibility that female anoles do not have female choice remains an open question. Despite these difficulties, more studies of female choice in relation to performance traits are clearly needed to understand the relative balance of male competition and female choice for the evolution of performance traits.

8.7 Asymmetry and performance

For bilaterally symmetrical organisms, a high level of symmetry is essential for animals to function properly. One attractive, yet controversial, hypothesis argues that asymmetries in different parts of an organism, such as between the left and right jawbones for example, arise as a result of environmental perturbations to organisms early in development ("fluctuating asymmetry"). Some work suggests that highly symmetrical individuals are more attractive to the opposite sex and more likely to survive compared to less symmetrical individuals. However, other researchers have found no relationship between levels of fluctuating asymmetry and any measure of fitness, performance, or mating success.

High levels of fluctuating asymmetry could influence whole-organism performance capacities in two potentially interrelated ways. First, individuals with high levels of asymmetry may be overall of lower "quality" compared to individuals with low levels of asymmetry. A second effect may be purely physical; when faced with highly physical tasks that require large amounts of coordination, such as fighting with a congener, highly asymmetric individuals may be at a decided disadvantage. Crabs offer an excellent example for testing this idea, as individuals fight strongly for access to territories that ultimately dictate reproductive success. Most crabs (fiddler crabs being an exception) fight with bilaterally symmetrical claws, and hence high levels of asymmetry could be influential in dictating fight performance.

In an examination of the shore crab *Carcinus maenas*, Sneddon and Swaddle (1999) found evidence for asymmetry playing a key role in settling agonistic disputes. These authors examined the relationship among trait asymmetry (for directional as well as fluctuating asymmetry) and the result of agonistic encounters among male shore crabs that were matched for size. They determined that the cheliped ("weapon claw") directional asymmetry was in fact not related to how fights ended but that the level of fluctuating asymmetry in the fifth pereiopod (one of the walking limbs) was negatively related to the probability of earning victory during conspecific encounters. This pattern was explained in the context of a functional advantage for symmetric individuals, as the fifth pereiopod is very important for maintaining stability and balance when these animals fight. Nevertheless, the authors did not find any evidence that asymmetry is a good overall metric of intrinsic individual quality.

A second with lacertid lizards found a somewhat different result. In the lizard *Lacerta monticola*, dominant males tend to also have relatively large heads, a trait which is known to be correlated with high bite forces. However, these same males also had more asymmetrical femurs and, overall, suffered reduced sprint speeds. Hence, dominance may come at a cost in that these dominant but asymmetrical males, while favored by sexual selection, may not be favored by natural selection (Lopez and Martin 2002). Despite the promise of these studies, several notes of caution are needed. First, studies of asymmetry are much like studies of natural selection—negative results are often unlikely to be published. Therefore, we may be only seeing interesting positive links between performance and asymmetry, without being aware of the myriad of negative data that have never been published.

References

Andersson M. 1994. Sexual Selection. Princeton University Press, Princeton, NJ.

Balmford A, Thomas ALR, Jones IL. 1993. Aerodynamics and the evolution of long tails in birds. Nature 361:628–631.

Brandt Y. 2003. Lizard threat display handicaps endurance. Proceedings of the Royal Society of London B 270:1061–1068.

Emlen DJ. 2014. Animal Weapons: The Evolution of Battle. Henry Holt and Company, New York, NY.

Evans JP, Kelley JL, Bisazza A, Finazzo E, Pilastro A. 2011. Sire attractiveness influences offspring performance in guppies. Proceedings of the Royal Society of London B 271:2035–2042.

Hunt J, Bussiere LC, Jennions MD, Brooks R. 2004. What is genetic quality? Trends in Ecology and Evolution 19:329–333.

Husak JF, Ribak G, Wilkinson GS, Swallow JG. 2011. Compensation for exaggerated eye stalks in stalk-eyed flies (Diopsidae). Functional Ecology 25:608–616.

Irschick DJ, Briffa M, Podos J. 2015. Animal Signaling and Function: An Integrative Approach. Wiley-Blackwell Press, Hoboken, NJ.

Kodric-Brown A, Nicoletto PF. 1993. The relationship between physical condition and social status in pupfish *Cyprinodon pecosensis*. Animal Behaviour 46:1234–1236.

Kodric-Brown A, Nicoletto PF. 2005. Courtship behaviour, swimming performance, and microhabitat use of Trinidadian guppies. Environmental Biology of Fishes 73:299–307.

Kotiaho JS. 2000. Testing the assumptions of conditional handicap theory: Costs and condition dependence of a sexually selected trait. Behavioral Ecology and Sociobiology 48:188–194.

Kotiaho JS, Alatalo RV, Mappes J, Nielsen MG, Parri S, Rivero A. 1998. Energetic costs of size and sexual signalling in a wolf spider. Proceedings of the Royal Society of London B 265:2203–2209.

Lailvaux SP, Hathaway J, Pomfret J, Knell R. 2005. Horn size predicts physical performance in the beetle *Euoniticellus intermedius* (Coleoptera: Scarabaeidae). Functional Ecology 19:632–639.

Lailvaux S, Herrel A, Vanhooydonck B, Meyers J, Irschick DJ. 2004. Performance capacity, fighting tactics, and the evolution of life-stage male morphs in the green anole Lizard (*Anolis carolinensis*). Proceedings of the Royal Society of London B 271:2501–2508.

Lailvaux S, Irschick DJ. 2006a. A functional perspective on sexual selection: Insights and future prospects. Animal Behaviour 72:263–273.

Lailvaux S, Irschick DJ. 2006b. No evidence for female association with high-performance males in the green anole lizard, *Anolis carolinensis*. Ethology 112:707–715.

Lailvaux S, Irschick DJ. 2007. The evolution of performance-based male fighting ability in Caribbean *Anolis* lizards. American Naturalist 170:573–586.

Lappin AK, Husak JF. 2005. Weapon performance, not size, determines mating success and potential reproductive output in the collared lizard (*Crotaphytus collaris*). American Naturalist 166:426–436.

Leal M. 1999. Honest signalling during prey–predator interactions in the lizard *Anolis cristatellus*. Animal Behaviour 58:521–526.

Lopez P, Martin J. 2002. Locomotor capacity and dominance in male lizards *Lacerta monticola*: A trade-off between survival and reproductive success? Biological Journal of the Linnean Society 77:201–209.

Maynard-Smith J, Harper D. 2003. Animal Signals. Oxford University Press, Oxford.

Moczek AP, Emlen DJ. 2000. Male horn dimorphism in the scarab beetle, *Onthophagus taurus*: Do alternative reproductive tactics favour alternative phenotypes? Animal Behaviour 59:459–466.

Prestwich KN, Walker TJ. 1981. Energetics of singing in crickets: Effect of temperature in three trilling species (Orthoptera: Gryllidae). Journal of Comparative Physiology 143:199–212.

Searcy WA, Nowicki S. 2005. The Evolution of Animal Communication: Reliability and Deception in Signaling Systems. Princeton University Press, Princeton, NJ.

Sinervo B, Lively CM. 1996. The rock–paper–scissors game and the evolution of alternative male strategies. Nature 380:240–243.

Sinervo B, Miles DB, Frankino WA, Klukowski M, DeNardo DF. 2000. Testosterone, endurance, and Darwinian fitness: Natural and sexual selection on the physiological bases of alternative male behaviors in side-blotched lizards. Hormones and Behavior 38:222–233.

Sneddon LU, Swaddle JP. 1999. Asymmetry and fighting performance in the shore crab, *Carcinus maenas*. Animal Behaviour 58:431–435.

Vanhooydonck B, Herrel A, Van Damme R, Irschick DJ. 2005. Does dewlap size predict male bite performance in Jamaican *Anolis* lizards? Functional Ecology 19:38–42.

Wells KD, Taigen TL. 1989. Calling energetics of a neotropical treefrog, *Hyla microcephala*. Behavioral Ecology and Sociobiology 25:13–22.

Wilson RS, Angilletta MJ Jr. 2015. Dishonest signaling during aggressive interactions: Theory and empirical evidence. In Irschick DJ, Briffa M, Podos J, eds. Animal Signaling and Function: An Integrative Approach. Wiley-Blackwell Press, Hoboken, NJ, pp. 205–228.

Wilson RS, Angilletta MJ Jr, James RS, Navas C, Seebacher F. 2007. Dishonest signals of strength in male slender crayfish (*Cherax dispar*) during agonistic encounters. American Naturalist 170:284–291.

Zahavi A, Avishag Z. 1997. The Handicap Principle: A Missing Piece of Darwin's Puzzle. Oxford University Press, Oxford.

9 Extreme performance
*The good, the bad, and
the extremely rapid*

9.1 Extreme performance: If the Olympics were open to oribatid mites

Though many of us don't care a whit about most Olympic sports most of the time, once they arrive in full every few years, we are often glued to the screen, in part because of patriotic sentiments but also because some of the sports are so outlandishly hard (ski jumping comes to mind). The sheer specialization and difficulty of these feats raises new possibilities for extreme human achievements. An example is the 1988 figure skating World Championships in Hungary, where Kurt Browning performed the first quadruple jump in competition, thus ushering in a new era of competition for figure skating! This level of "extreme" performance isn't limited to the Olympics: one does not have to look far for other examples such as "extreme" cooking, "extreme" dinosaurs, or "extreme" animal encounters (e.g., tarantula vs. rattlesnake)?

Comic books are renowned for their creation of outlandishly muscled heroes that signal otherworldly strength. However, the saying that truth is stranger than fiction is perhaps never more valid than in the context of the extreme athletic achievements of animals. If one were to hold a combined human-animal Olympics, the results would not even be close. Let us first consider strength. Hossein Rezazadeh (from Iran) holds one of the top weightlifting efforts of 263 kg, which is about twice his body weight. In contrast, an oribatid mite has been documented lifting 1,180 times its weight. That is equivalent to Hossein lifting an adult blue whale, the largest known animal ever to have existed. The mite wins!

Animal Athletes. Duncan J. Irschick and Timothy E. Higham. © Duncan J. Irschick and Timothy E. Higham 2016. Published in 2016 by Oxford University Press.

The human-animal comparisons can be made in almost all categories. Michael Phelps is an incredibly fast swimmer, reaching approximately 10 km/h. A bottlenose dolphin, by comparison, can reach approximately 35 km/h, taking the gold medal from Phelps. Usain Bolt, although very fast, would have little chance against a cheetah. However, they are not as far off as you might think, with the fastest cheetah beating Bolt in the 100 m sprint by only 3.63 s. In addition to maximum speed and strength, one could look at endurance performance. The recent, and incredible, accomplishment of Diana Nyad comes to mind. She became the first person (without a shark cage) to swim from Cuba to Florida. She swam approximately 160 km in almost 53 h. In contrast, polar bears have been documented swimming continuously for almost 700 km over a period of 9 days (232 h). All of this from an animal that spends most of its time on land! These examples clearly show that humans are not superior in a specific category of performance. That said, let's keep in mind that humans compete in all of these different categories and can, therefore, be considered generalists. Because of trade-offs, there are not many animals that can perform a wide range of behaviors at a maximum level. Thus, comparing humans and other animals should be taken with a grain of salt.

9.2 Extreme performance: Overcoming limits

While it's fun to ponder hypothetical match-ups between humans and nonhuman animals, there is a more intellectually rewarding avenue for studying extreme performance in animals, namely, as a way to understand the process of adaptation through natural selection. If one flips open any textbook in animal physiology, it's often the extremes that are the focus of each chapter. A great deal of what we understand about thermal adaptation comes from studies of animals that live in extremely cold (i.e., the Antarctic) or hot (deserts) environments. Likewise, this logic is repeated for studies of osmoregulation (the maintenance of the osmotic pressure of an organism's fluids), locomotion, the nervous system, and so on. In other words, studying extremes is useful because it brings into relief the mechanisms that animals use to cope with environmental variation and, by studying the extremes, we may gain insight into how all organisms work.

It's not hard to find some striking extremes that defy a simple explanation. Elephant seals can dive up to 1,500 meters below the surface, and their heart rate slows to only a few beats per minute. Peregrine falcons can streak through the air at speeds up to 97 m/s (that's 350 km/h!) to capture prey. Tiny snapping mantis shrimp can accelerate striking appendages at accelerations of over 65–104 km/s^2, crushing hard prey in milliseconds! In all of these and other cases, extreme abilities have evolved to fulfill some important ecological need. Elephant seals dive deeply to capture prey while minimizing the threat of predators such as white sharks. Peregrine falcons use their amazing speeds to capture prey, and snapping shrimp use their appendages to break open hard shells.

One of the emergent principles is that animals often have behavioral, functional, or morphological "tricks" that enable them to achieve far higher levels of performance than might be apparent from a simple inspection of their physical form (Higham and Irschick 2013). In some cases, animals achieve extreme performance abilities through the possession of novel anatomical or physiological structures or mechanisms, or some combination thereof. Most lizards have poor endurance capacities compared to mammals or birds but varanid lizards (e.g., the Komodo dragon) have enhanced their aerobic capacity through the presence of a gular pump in their throat that acts much like a backup air pump (Owerkowicz et al. 1999). Some arctic fish can tolerate extremely cold water through the use of "antifreeze" molecules.

Before moving forward, we should address the concept of "limit." The limit for a particular performance will depend on the mechanisms underlying that performance but can be defined as the "traditional" rules that dictate how a physiological or functional process works. These "rules" are typically constructed based on models for how different structures work but, as can be seen in this chapter, rules are indeed made to be broken. An example is useful here. For jumping, which is a rapid and powerful movement, it seems obvious that muscle performance will be limiting. Muscle power is the force of the muscle multiplied by how quickly it can contract, and this factor, in turn, determines how fast an animal can jump. However, the spectacular jumping ability of frogs hints that there may be alternative mechanisms for limiting performance (Astley et al. 2013). With some previous evidence that frogs use an elastic mechanism to amplify power output during jumping, Thomas Roberts and colleagues examined peak muscle power output for the plantaris muscle in three species of frogs—Cuban tree frogs, *Osteopilus septentrionalis*, leopard frogs, *Rana pipiens*, and cane toads, *Bufo marinus*—and compared it to the power produced during jumping, using force plate ergometry (Roberts et al. 2011). Although not the case for the cane toads, the ability to stretch tendons and release elastic energy, much like what we do with elastic bands or slingshots, allows energy to be stored slowly and then released rapidly in the tendons of Cuban tree frogs and leopard frogs. In fact, the release of energy from tendons, resulting in the movement of the frog, can occur much faster than a muscle could contract. Thus, this behavior is no longer limited by muscle, because of this intriguing way of storing and releasing energy. Circumventing the limits of muscle physiology in this manner is surprisingly common and can be observed in a variety of animals from seahorses to turkeys (Higham and Irschick 2013).

Therefore, a performance enhancer is any morphological, physiological trait (or substance) that overcomes a previously defined limit based on existing

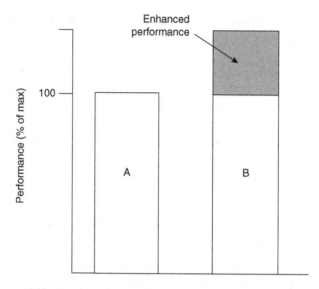

Fig. 9.1 The potential increase in performance, beyond physiological limits, due to performance enhancers. Many animals circumvent physiological limits by using morphological enhancers, such as a tendon coupled with a catch mechanism. Graph taken from Higham and Irschick (2013) with permission from Springer.

models of the way a given set of structures should function (fig. 9.1). In terms of morphology, this could be a tendon that can store and release energy during jumping in a frog, therefore bypassing the limits imposed by muscle physiology. Physiological enhancers could be in the form of hormones (e.g., testosterone, a form of performance-enhancing drug [PED]), which appear on the news every year in relation to many sports such as cycling, football (American), and baseball. PEDs are a hot topic these days, with the admission of drug use by Lance Armstrong and others but also because it is becoming evident that PEDs are used in a much wider range of human *and* nonhuman animal sports than previously thought. Performance enhancers should not be confused with simple modifications to behavior or efficiency of motion. Olympic jumpers used to jump over the bar belly first until Dick Fosbury pioneered the "back-first" "Fosbury flop" that is widely used today. This change in jumping behavior resulted in a shift of maximum jumping performance (jump height). In nature, with all else being equal, one might see some species utilizing a way of jumping (behavior) that allows them to exceed the performance of another species.

9.3 Need for speed: Extremely rapid movements

Rapid movements of the whole body, or of individual body parts, can form some of the most eye-catching highlight reels produced by evolution. Much of the variation in how fast animals can move is a function of body size (Mc-Mahon 1975). Small animals are capable of very rapid short-term and short-distance movements (particularly high accelerations); therefore, small animals, such as fruit flies or fleas, can achieve much higher relative velocities (velocity divided by body size) than larger animals. On an absolute scale, however, larger animals will produce higher velocities over longer distances, as can be observed from the average speed of a horse versus a mouse over a kilometer, for example. One of the reasons that larger animals cannot achieve extremely high speeds is the influence of gravity. Because larger animals have a larger mass, they also must work harder against gravity compared to small animals, especially during flight. This is one of the reasons that a fruit fly can twirl around your nose at high speeds whereas a large swan can barely get off the ground.

Because of the importance of effectively capturing prey, some animals have evolved extremely rapid tongue projection movements for capturing elusive prey. A well-known example of fast movement is the projectile tongue of chameleons. Chameleons are very slow, moving in mostly arboreal habitats with very deliberate motions. Because of this, a mechanism to catch quickly moving prey is needed, which is where tongue projection gives them an advantage. In addition, projecting a small structure toward a moving prey item is more likely to go unnoticed than an entire animal moving quickly toward the prey. Therefore, moving slowly and projecting the tongue might actually be more effective than being a quickly moving predator. The unique protrusion mechanism of chameleons involves an accelerated lengthening of the tongue and the muscle in the tongue, resulting in incredibly high power output (Higham and Anderson 2013). Tongue projection is also an example of power amplification, given that the muscles alone are not able to provide the power observed. Interestingly, temperature also has little impact on chameleon tongue projection (fig. 9.2; Anderson and Deban 2010), a rare result given that temperature often impacts biomechanics.

In addition to chameleons, there are many frog species that utilize rapid (as low as 30 ms in duration) projections of their tongue to obtain prey. In the majority of animals that use tongue projection to capture prey, tongue movement is created through active muscle recruitment, and the length of tongue projection is small (<2% of body length or less). Some salamanders, however, such as those in the genus *Hydromantes*, can project their tongues outward up to 100% of body length across only a few milliseconds (Deban et al. 1997; Deban and Dicke 1999). Examination of the anatomy surrounding their tongues reveal muscle fibers that constrict the tongue, thereby projecting it from the salamander's mouth at high velocities, in the same way that a wet watermelon seed can be squeezed between fingers! Retractor muscles then "reel in" the elongated tongue after usage. The end result is a rapid ballistic projection in which the tongue moves without additional energetic input beyond the initial squeezing. The ecological value of rapid tongue projection seems straightforward. By using rapid tongue

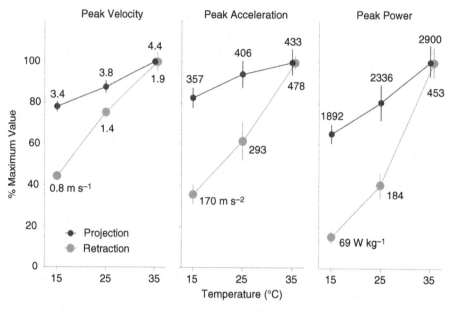

Fig. 9.2 Graph showing the limited impact of temperature on the velocity, acceleration, and power of tongue projection (small, dark gray circles), but strong impact on tongue retraction (large, light gray circles) in chameleons. Tongue projection is powered by elastic recoil, whereas tongue retraction is powered by regular muscle contractions. Graph taken from Anderson and Deban (2010) with permission from the National Academy of Sciences.

movements as opposed to rapid locomotion, salamanders and chameleons can capture relatively large prey with little extra investment of energy.

For flying or swimming animals, fluid dynamics also plays a role. Air and water both act like fluids, albeit with very different properties. In this regard, extremely small animals live in a world dominated by viscous forces, and locomotion in water for them is challenging. Imagine swimming through a pool full of maple syrup! For much larger animals, such as dolphins, inertial forces dominate, and streamlined body shapes are much more beneficial. In the terrestrial world, another reason why larger animals cannot achieve velocities or accelerations as high as small animals can is due to the total amount of power produced by muscle or tissues relative to body mass. Small animals generally have very high ratios of power to mass, whereas larger animals have lower ratios. Just think of the variety of cars on the market—a Porsche may exceed 500 horsepower and have a mass of 1,500 kg. In contrast, a Hummer H2 has a power output of approximately 400 horsepower but has a mass of 3,000 kg. This gives a power-to-mass ratio of 0.33 for the Porsche, and 0.13 for the Hummer. This 20% difference in horsepower actually represents a 40% decrease in the power-to-mass ratio. Now, compare this to a 1963 Volkswagen Beetle (proudly owned by one of the authors, Tim Higham), which comes in with a whopping

40 horsepower and has a mass of 725 kg. This power-to-mass ratio is 0.06, 2 times lower than the Hummer and 6 times lower than the Porsche. (This power-to-mass ratio, of course, does not apply to Herbie!) In the same way that cars differ in their ratios of power to mass, smaller animals can generally perform much more athletic feats than larger animals, at least during terrestrial or aerial movements (as noted above, movement in water and other fluids is more complex).

While much of the discussion of extreme locomotor abilities centers on charismatic animals such as cheetahs or horses, some of the animals that can move the fastest, at least over short distances, are far smaller and less well known. Sheila Patek and her colleagues have performed experiments with two groups of animals—mantis shrimp and trap-jaw ants—both of which possess remarkable abilities to move appendages at extremely high velocities and forces. One of the lessons from both species is that extreme performance capacities appear to have evolved for multiple functions (fig. 9.3). Mantis shrimp are a diverse group of small (~1–7 cm long) invertebrates that use an external appendage (mouthpart) to damage, shock, or kill prey. They also use these appendages to

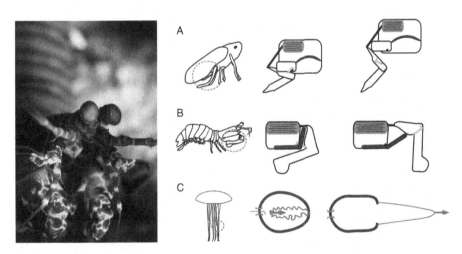

Fig. 9.3 *Left*, An image of a mantis shrimp. *Right* (*A–C*), Figure depicting the different mechanisms of elastic energy storage and recoil. Rapid movements can be driven by different structures, such as those in froghoppers, mantis shrimp, and cnidarians. *A*, Froghoppers can jump through the preloading of an elastic pleural aric that is situated between the body and the femur. The tension can then be released via a latch-like system. *B*, Mantis shrimp generate powerful strikes with a raptorial appendage that employs an elastic ventral bar. This mechanism is analogous to that of a tape spring that is released using a latch. *C*, Some cnidarians sting using their poisonous stylets which can be fired from pressurized cellular capsules. For all of these images, the engine is depicted as the medium gray rounded rectangle, the spring is depicted as dark gray, and the latch (when present) is light gray. The figure on the right is taken from Patek et al. (2011) with permission from the Company of Biologists Ltd. The image of the mantis shrimp on the left is from Wikimedia Commons.

contest and defend their small burrows from potential rivals. Mantis shrimp will unfurl their appendages at extremely high velocities (up to 29 m/ s, or 64 mph), and one can hear the "pop" from the ensuing cavitation bubble popping in the water (fig. 9.3; Patek et al. 2011; Cox et al. 2014)! Indeed, the energy from the bubble (which reaches temperatures of about 7,000°C) forming and bursting provides enhanced force beyond that of the initial strike (Patek and Caldwell 2005). In other words, the shrimp achieves two strikes in one. This remarkable performance allows these shrimp to produce forces strong enough to break open the shells of mollusks. A second example from the research done by Sheila Patek and her collaborators (2006) comes from the study of ant trap jaws, which have evolved independently several times. The trap jaws have oversized mandibles that, when cocked, lie at nearly a 180° angle and will close forcefully when released. The most obvious function of these jaws is in defense, as the jaws can snap shut painfully on an intruder to a nest, or on a potential predator, but they can also be used to capture prey. A less obvious function is escape from predators. When the jaws are snapped shut on hard surfaces, the force of the movement can propel the ant into the air (fig. 9.4)! This multifunctional value of the trap jaws may explain why such an unusual specialization has arisen several times during the course of evolution (Patek et al. 2006). The closing abilities of the trap jaws are impressive, as snaps can take less than 2.5 ms, with velocities of between 35 and 64 m/s, and accelerations of 10^5 times the force of gravity. The trap-jaw mechanism is similar to that of a catapult: the mandibles are slowly opened and kept in a locked and wide position until a trigger hair that is innervated by several large neurons is stimulated, after which the mandibles are freed from the catch and rapidly close (fig. 9.5; Gronenberg 1996). Comparisons of closely related trap-jaw ant

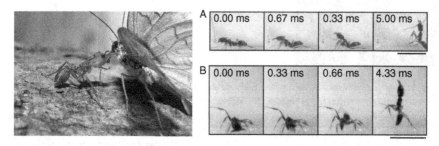

Fig. 9.4 *Left,* Image of a trap-jaw ant, *Odontomachus bauri. Right (A, B),* Images of a bounder defense maneuver and an escape jump. *A,* In the bounder defense maneuver, the ant approaches the intruder with its jaws open and cocked. The jaws close rapidly against the intruder, and this action propels the ant's head and body away and upward. *B,* The escape jump involves a strike that propels the animal upward. Images on the right are taken from Patek et al. (2006) with permission from the National Academy of Sciences. The image of the trap-jaw ant on the left is from Wikimedia Commons.

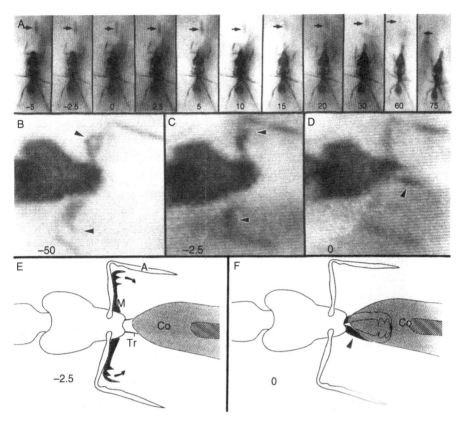

Fig. 9.5 *A*, Images from a high-speed (400 Hz) series demonstrating the mandible strike of a trap-jaw ant (*Strumigenys* sp.). The time (in milliseconds) is shown at the bottom of each image, and the images were taken both before and after the strike. The strike occurs at time 0. The arrows indicate the gut of the insect prey. Note that the insect prey eludes the strike. *B–D*, Close up of the mandibles striking; the mandibles are indicated with arrowheads. See Gronenberg (1996) for more details. Taken from Gronenberg (1996) with permission from the Company of Biologists Ltd.

species with heads of different sizes show that the total amount of force that can be produced by the jaws is closely related to the size of the head, with ant species that possess large heads being able to produce higher forces than ant species with small heads can.

Small invertebrates are also some of the most impressive jumpers for their body size. While a flea or other similarly sized insects cannot jump as far in absolute terms as a human or frog, when considered as a proportion of body size, their performance is remarkable (Burrows 2003). Their small size provides both detriments and benefits in terms of jumping. On the plus side, being small means that they can accelerate extremely quickly because they don't have to work as hard against gravity as a larger animal would. For this reason, jumping accelerations are much higher in very small insects compared to larger animals. However, because their limbs are so small, insects can only accelerate over a short distance before they limbs leave the ground. Thus, there is a trade-off among animals of different sizes between jumping acceleration and the time over which acceleration occurs. As size increases, this trade-off eventually works in favor of larger animals, as Carl Lewis could jump father than any flea, but a closer inspection of the jumping data reveals an interesting story. The diminutive froghopper insect weighs about 12 mg and can complete a jump in about 1 ms, with accelerations between 2,800 and 4,000 m/s^2. By comparison, the duration of a standing jump from a human takes approximately 0.8 s, with accelerations of 20 m/s^2. Even for its small size, the accelerations of the froghopper during jumping are remarkable and beg the question of how they can be achieved. Both froghoppers and fleas use a catapult mechanism to enhance their jumping capacities. Potential energy is stored slowly through great force and leverage, until it is propositioned to be released, at which point, very large amounts of stored energy are released quickly, thus powering the jump.

In terms of locomotion over longer distances, larger animals tend to be the most impressive performers. The peregrine falcon, *Falco peregrinus*, is considered to be the fastest bird on earth and can use its speed to capture prey in midair (Tucker 1998; Tucker et al. 1998). They can achieve diving speeds around 150 m/s, or about 335 mph, although these values are rough estimates. From a practical perspective, how do these birds achieve such high speeds in midair? The answer seems to be that they need a lot of air space to do so. It's very hard to study this kind of rapid diving in nature, but simple biomechanical models can yield some predictions. One factor that enhances speed is the size of the bird. The larger the bird, the faster they can stoop since, unlike normal flight, the bird is diving, and larger animals will dive faster because the force of gravity works in their direction. These birds would likely have to dive over long distances, such as up to 1,000 m, to achieve top speeds and, depending on their angle of diving, would lose between 300 and 1,000 m in altitude. These large numbers suggest that it is likely rare that falcons would achieve such high speeds in nature, given the great effort they would have to expend both to reach top speeds and then slow down. Thus, while animals may sometimes employ their maximum performance capacities in nature, there are good reasons why they do not use them all of the time.

9.4 The evolution of morphological novelties

The ability to perform exceptional feats of athleticism is sometimes due to the unique evolution of a specialized structure (Muller and Wagner 1991). One example of a radical modification of an existing structure is the pharyngeal jaw of moray eels. Anyone who has watched the *Alien* movie series is likely fascinated and terrified by the raptorial and miniature jaws that erupt from the large mouth of the alien, thus killing the hapless victim in a satisfyingly horrific manner. It turns out a less spectacular (though highly effective) version of these jaws exist in nature! Moray eels are some of the most fearsome apex predators in coral reefs and are known to feed on cephalopods or fishes that are sometimes larger than they are. Most large fishes other than sharks rely on suction to pull prey into their oral cavities, but the suction capacity of moray eels is limited. These eels instead grab prey with the sharp teeth of their primary jaw and then use pharyngeal jaws, which are normally hidden deep in the mouth, to grab the prey and pull it down the throat and into the digestive system (fig. 9.6; Mehta and Wainwright 2007). Further, moray eels have undergone many evolutionary modifications to their oral anatomy to enable this remarkable structure to function. This structure can exist as a result of the elongation of the muscles that control the jaws, as well as a reduction of nearby gill-arch structures. Many other fishes possess pharyngeal jaws, which serve a similar purpose in prey transport, although not in such a spectacular fashion. Moray eels also vary in the size of their pharyngeal jaws, with those species that feed on a larger selection of food items possessing a more elaborate and developed set, while more specialized moray eels possess less developed jaw characteristics.

Novel modifications of morphology are especially useful for animals that reside in physically challenging habitats. Burrowing is especially challenging because of the great effort involved (try burrowing yourself, and see how far

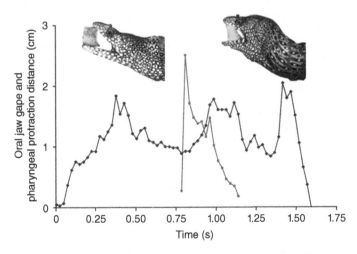

Fig. 9.6 Graph showing the protraction of the pharyngeal jaws (light gray trace) following peak oral gape (dark gray trace) in a moray eel, *Muraena retifera*. Recurved teeth on the pharyngeal jaws grip the prey, allowing the fish to retract the prey towards the esophagus. Graph taken from Mehta and Wainwright (2007) with permission from Nature Publishing group.

you get!) and the low level of oxygen in fossorial environments. While such conditions might not pose a problem for animals that do not require oxygen, or are not in a hurry, for animals that need to burrow relatively quickly to capture prey or to perform other tasks, performing proficiently poses a mechanical challenge. Caecilians are among the most remarkable burrowers on earth. They are an ancient group of burrowing amphibians that largely reside in tropical soils in New and Old World habitats. These animals burrow using powerful movements of their snout against the walls of the tunnel. Jim O'Reilly and several colleagues have performed elegant experiments to determine how these animals can achieve their remarkable burrowing performance. They placed small pressure transducers into the internal body cavity of a caecilian species (the Central American *Dermophis mexicanus)* and found that these animals power locomotion hydrostatically through a crosshelical array of tendons surrounding their body cavity (O'Reilly et al. 1997). Loose folds of skin that surround the body, as well as the lack of connections between their skin and other structures, enable the caecilians to move their vertebral column independently of their skin, thus freeing them to expand their bodies hydrostatically. This unique morphological arrangement and functional ability results in burrowing forces of over 20 N (remember, a newton is about the weight of a small apple, hence the name), which is about twice as great as the force generated by burrowing snakes that do not possess this unique morphological arrangement.

9.5 Morphological and physiological mechanisms of extreme performance

The evolution of morphological novelties has allowed some animals to achieve high levels of performance; but, in many other cases, there are deeper alterations in physiology and structure that require a closer look. High-frequency movements of various body parts pose a direct challenge to classic ideas about how muscles work, especially if this trend occurs over longer time periods. Several kinds of animals are capable of extremely high-frequency movements; for example, toadfish emit a "boatwhistle call" at around 200 Hz, and rattlesnakes shake their tails at around 90 Hz (fig. 9.7; Rome et al. 1996). It's tempting to think that nature has already provided a solution to this problem because of the presence of different muscle fiber types that are each uniquely suited for different levels of force production and contraction speeds. Slow-twitch muscle fibers enable less powerful but more sustainable movements (e.g., walking) whereas fast-twitch muscles enable animals to produce forceful but less sustainable movements over short time periods (e.g., jumping). Nevertheless, typical fast-twitch muscle fibers would be severely challenged to pull off the high-frequency movements seen in toadfish, especially over

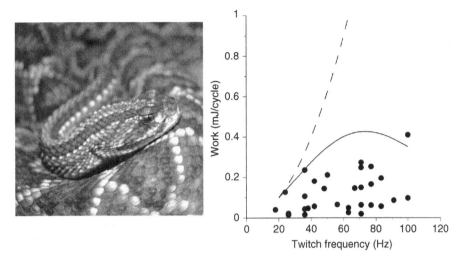

Fig. 9.7 *Left*, An image of the western diamondback rattlesnake, *Crotalus atrox*. *Right*, Graph showing the mechanical work associated with tail rattling. The circles represent data collected from rattlesnakes, and the dashed line indicates the hypothetical work, based on a pendulum model, that would be required if displacement were maintained as frequency increased. The solid line represents a model in which pendulum displacement decreases as frequency increases. Graph taken from Moon et al. (2002) with permission from the Company of Biologists Ltd. Image of the western diamondback rattlesnake is taken from Wikimedia Commons.

long time periods. In addition to needing a high level of coordination between nerves, muscles, and the brain, these high-frequency movements are challenging because of basic rules about how muscles work. Further, muscles need to relax prior to performing additional contractions, but relaxation takes time, posing yet another problem.

Let's focus first on the toadfish. In order to understand how this animal can produce such a high-frequency call, we need to appreciate some of the basic principles of muscle mechanics. Muscle fibers produce force by means of attachments between myosin heads and actin filaments. As muscles contract, these "rows" become shorter and pull the muscle fibers across one another in much the same way as an oar pushes a canoe across the water. This "sliding filament mechanism" is the most accepted model for how muscles work. However, a limiting step in this process is the ability of the sarcoplasmic reticulum to first secrete calcium, which initiates muscle contraction, and then to resequester it, thus causing muscle relaxation. A notable feature of the muscles found in sound-producing toadfish is their ability to release and resequester calcium from the sarcoplasmic reticulum more rapidly than a more typical muscle can (fig. 9.8; Rome et al. 1996). The end result is the generation of extremely rapid muscle contractions, thus facilitating the boatwhistle. Other specializations also appear to play a role. The sound-producing muscles of toadfish also

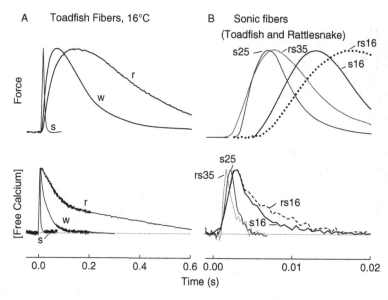

Fig. 9.8 *A*, Function in toadfish muscle fibers, including slow-twitch red fibers (r), fast-twitch white fibers (w), and superfast-twitch swim-bladder fibers (s). *B*, Function of sonic fibers from the rattle-snake tail shaker muscle at 16°C (rs16) and 35°C (rs35), and fibers from the toadfish swim bladder at 16°C (s16) and 25°C (s25). In both *A* and *B*, the upper panels indicate muscle force, and the lower panels show the calcium transients. The average half-widths of the calcium transients were over 100 ms for red muscle and less than 5 ms for superfast muscle. Redrawn from Rome et al. (1996) with permission from the National Academy of Sciences.

exhibit an extremely fast crossbridge detachment rate and a fast kinetic rate of calcium detachment from the crossbridges. Remarkably, the muscle fibers used to power the rattle of rattlesnakes seems to exhibit many of the same properties, thus allowing them to rattle at very high frequencies. This pattern of convergent evolution reveals how natural selection for rapid movements can select for distinctive properties of normally highly conserved muscle physiology. The rattle of rattlesnakes is especially perplexing, as they can rattle at high frequencies for long periods of time (over an hour) while expending relatively little energy (Moon et al. 2002). Further, muscles involved in rattling show an interesting property. As the force of rattling increases, the energetic cost of muscular twitches remains constant. The rattler muscles seem to show a unique energy saving property. The ecological significance of these unique muscle properties is obvious; when encountered with a threat, rattlesnakes have the capacity to provide a warning signal that is long lasting and relatively energetically inexpensive.

To contract this quickly, something has to give. Although they need to contract quickly to successfully perform the behavior, toadfish swim bladders and rattlesnake tails don't need to generate a lot of force. This is a good thing because there simply isn't room in a muscle cell to fit everything in! An ultrafast muscle that contracts for a long period of time needs oxygen, and lots of it. Mitochondria are key here, so muscles that contract at extreme speeds are packed with mitochondria. Thus, space for myosin and actin, the force-generating components of muscles, is limited. Although you could test this lack of force by grabbing the tail of a rattling rattlesnake, we would advise that you leave that approach to the professionals!

Animals can also evolve interesting material properties that can enable novel kinds of performance. Male hummingbirds will perform extreme locomotor maneuvers as a signal to onlooking females, ideally impressing females with sounds and impressive displays of speed. Christopher Clark recently performed eloquent experiments to figure out *how* the feathers of a humming-bird produce sound, and he did this by putting individual feathers in a wind tunnel (Clark et al. 2011). When the wind speeds in the tunnel reached speeds exhibited by male birds during dives, the feathers emitted loud noises when they vibrated. It turns out that the properties of feathers that are important for flight also make them good at producing sound—the feathers are stiff but also elastic. It is also apparent that different species have different feather properties, and therefore make different sounds when exposed to the same speed.

9.6 Behavioral means for greatly enhancing performance

Based purely on biomechanical principles, one might expect linear relationships between morphology, performance, and fitness within species, but such strict relationships are not always observed. Part of the reason is that organisms often have the ability to circumvent mechanical constraints by employing novel behaviors. The evolution of these novel behaviors can open up new evolutionary pathways that might normally never be employed.

Early research on animal performance emphasized the inextricable link between morphological structure and performance and paid less attention to behavior as a potential factor that might either enhance or diminish performance. However, researchers now realize that, while the direct relationship between morphological structure and performance is often strong, animals have behavioral tricks for enhancing performance to match the specific demands. Two examples of this are sidewinding in vipers and brachiation in monkeys. Lateral undulation is a common mode of progression in snakes, but this approach becomes less effective when moving on soft sand, such as in the deserts of Namibia or Southern California. Sidewinding is a mode of locomotion that minimizes the shear forces at contact by lifting some body segments while others remain in static contact with the ground. Thus, sidewinding has been proposed as an effective way of moving on yielding surfaces (Mosauer

1930; Marvi et al. 2014). This behavioral shift doesn't necessarily enhance performance but does allow the snake to maintain performance on an otherwise inaccessible substrate. Brachiation is another such modification in behavior found in primates. Brachiation involves swinging from branch to branch by one's forelimbs, always keeping the mass of the body underneath the point of contact. The next time you visit a zoo, look closely at the gibbons; they will likely be exhibiting brachiation as they move around. This form of locomotion mimics the movement of a pendulum and ultimately leads to the forelimbs being loaded in tension (Swartz et al. 1989).

Lizards have also modified their behavior to move across a substrate (water) that would seem inaccessible to them. Few vertebrates are capable of running across open water, but one kind of lizard, namely the basilisk, genus *Basiliscus*, is a documented water runner. This species has a tremendous size range, with relatively small juveniles (2–10 g) but quite large adults (>200 g in mass). Only juveniles and subadults are capable of running across water; thus, part of the secret to their water-walking success is body size. Water poses special problems for movement because, unlike materials that are at least semisolid, water yields to movement, preventing the active application of force that would enable animals to push forward. A daunting problem for any investigation of how basilisks can move across water is understanding the relevant forces involved. On solid surfaces, one can readily measure force by means of force platforms, which measure even extremely minute forces. Hsieh and Lauder (2004) used an ingenious method to quantify the kinematics of movement, as well as the forces used, in juvenile basilisks, *Basiliscus plumifrons*, as they moved across water. The researchers set up two perpendicular sheets of microscopic reflective particles that were illuminated by lasers (fig. 9.9). This method, known as digital particle image velocimetry, enables one to calculate force vectors created in moving fluids by dynamic actions, such as the movement of a limb across/through the water. The results showed that the lizards are able to generate enough support force for them to run across water and also that new kinematic strategies are needed for them to move across a yielding surface (Hsieh and Lauder 2004). Smaller juvenile basilisk lizards produced the highest values of support and propulsive forces during the first half of their steps, when the foot moves mostly vertically downward into the water. These juvenile lizards also produced high transverse reaction forces that changed from medial (79% body weight) to lateral (37% body weight) across the step. These forces may act to dynamically stabilize the lizards during water running.

While it is the large animals that often catch our eye, it is small animals that often exhibit the most impressive performance abilities, and nowhere is this more evident than in the realm of flight. Most of us have attempted to swat a fly, with great frustration, and the fruit fly *Drosophila* has served as a model system for understanding motor control of flight (Fry et al. 2003). The common

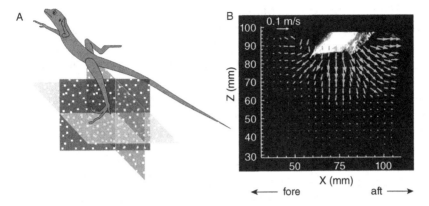

Fig. 9.9 Figure showing how basilisk lizards, genus *Basiliscus*, run on water. *A*, Schematic diagram of the experimental setup. *B*, Graph showing the velocity of fluid movement during the initial slap phase. Particle speed and direction are represented by vector lengths. The lizard's left foot produced the fluid motion in all panels. Graph taken from Hsieh and Lauder (2004) with permission from the National Academy of Sciences.

fruit fly is a remarkable flying creature that is powered by a pair of independently moving wings powered by several muscles that allow the wings to move up or down, side to side, or forward and backward (fore and aft). The ability to move their wings in multiple directions simultaneously is essential for these flies to perform their remarkable aerial feats, such as rapidly flying upside down. As any individual who flies regularly knows, human-built planes must obey certain aerodynamic rules or else fall out of the sky. In particular, with some exceptions, human-built planes are severely limited in their ability to maneuver in midair because of the risk of stall, which is the loss of lift caused by the separation of the air flow from the leading edge of the flying surface. By contrast, fruit flies appear to suffer little risk of catastrophic stall, in part because their extremely small size means that they have to expend far less energy to make rapid movements compared to large animals. Second, whereas human planes or helicopters use simple and well-understood flying surfaces, the wings of fruit flies are used in a far more complex fashion, often rotating in multiple directions during flight.

References

Anderson CV, Deban SM. 2010. Ballistic tongue projection in chameleons maintains high performance at low temperature. Proceedings of the National Academy of Sciences 107:5495–5499.

Astley HC, Abbott EM, Azizi E, Marsh RI, Roberts TJ. 2013. Chasing maximal performance: A cautionary tale from the celebrated jumping frogs of Calaveras County. Journal of Experimental Biology 216:3947–3953.

Burrows M. 2003. Biomechanics: Froghopper insects leap to new heights. Nature 424:509

Clark CJ, Elias DO, Prum RO. 2011. Aeroelastic flutter produces hummingbird feather songs. Science 333:1430–1433.

Cox SM, Schmidt D, Modarres-Sadeghi Y, Patek SN. 2014. A physical model of the extreme mantis shrimp strike: Kinematics and cavitation of Ninjabot. Bioinspiration and Biomimetics 9: 016014.

Deban SM, Wake DB, Roth G. 1997. Salamander with a ballistic tongue. Nature 389:27–28.

Deban SM, Dicke U. 1999. Motor control of tongue movement during prey capture in plethodontid salamanders. Journal of Experimental Biology 202:3699–3714.

Fry SN, Sayaman R, Dickinson MH. 2003. The aerodynamics of free-flight maneuvers in *Drosophila*. Science 300:495–498.

Gronenberg W. 1996. The trap-jaw mechanism in the dacetine ants *Daceton armigerum* and *Strumigenys* sp. Journal of Experimental Biology 199:2021–2033.

Higham TE, Anderson CV. 2013. Function and adaptation in chameleons. In Tolley KA, Herrel A., eds. The Biology of Chameleons. University of California Press, Oakland, CA, pp. 63–83.

Higham TE, Irschick DJ. 2013. Springs, steroids, and slingshots: The roles of enhancers and constraints in animal movement. Journal of Comparative Physiology B 183:583–595.

Hsieh ST, Lauder GV. 2004. Running on water: Three-dimensional force generation by basilisk lizards. Proceedings of the National Academy of Sciences 101:16784–16788.

Marvi H, Gong C, Gravish N, Astley H, Travers M, Hatton RL, Mendelson JR, Choset H, Hu DL, Goldman DI. 2014. Sidewinding with minimal slip: Snake and robot ascent of sandy slopes. Nature 346:224–229.

McMahon TA. 1975. Using body size to understand the structural design of animals: Quadrupedal locomotion. Journal of Applied Physiology 39:619–627.

Mehta RS, Wainwright PC. 2007. Raptorial jaws in the throat help moray eels swallow large prey. Nature 449:79–82.

Moon BR, Hopp J, Conley KE. 2002. Mechanical trade-offs explain how performance increases without increasing cost in rattlesnake tailshaker muscle. Journal of Experimental Biology 205:667–675.

Mosauer W. 1930. A note on the sidewinding locomotion of snakes. American Naturalist 64:179–183.

Muller GB, Wagner GP. 1991. Novelty in evolution: Restructuring the concept. Annual Review of Ecology and Systematics 22:229–256.

O'Reilly JC, Ritter DA, Carrier DR. 1997. Hydrostatic locomotion in a limbless tetrapod. Nature 386:269–272.

Owerkowicz T, Farmer CG, Hicks JW, Brainerd EL. 1999. Contribution of gular pumping to lung ventilation in monitor lizards. Science 284:1661–1663.

Patek SN, Baio JE, Fisher BL, Suarez AV. 2006. Multifunctionality and mechanical origins: Ballistic jaw propulsion in trap-jaw ants. Proceedings of the National Academy of Sciences 103:12787–12792.

Patek SN, Caldwell RL. 2005. Extreme impact and cavitation forces of a biological hammer: Strike forces of the peacock mantis shrimp (*Odontodactylus scyllarus*). Journal of Experimental Biology 208: 3655–3664.

Patek SN, Dudek DM, Rosario MV. 2011. From bouncy legs to poisoned arrows: Elastic movements in invertebrates. Journal of Experimental Biology 214:1973–1980.

Roberts TJ, Abbott EM, Azizi E. 2011. The weak link: Do muscle properties determine locomotor performance in frogs? Philosophical Transactions of the Royal Society B 366:1488–1495.

Rome LC, Syme DA, Hollingworth S, Lindstedt SL, Baylor SM. 1996. The whistle and the rattle: The design of sound producing muscles. Proceedings of the National Academy of Sciences 93:8095–8100.

Swartz SM, Bertram JEA, Biewener AA. 1989. Telemetered in vivo strain analysis of locomotor mechanics of brachiating gibbons. Nature 342:270–272.

Tucker VA. 1998. Gliding flight: Speed and acceleration of ideal falcons during diving and pull out. Journal of Experimental Biology 201:403–414

Tucker VA, Cade TJ, Tucker AE. 1998. Diving speeds and angles of a gyrfalcon (*Falco rusticolus*). Journal of Experimental Biology 201:2061–2070.

10 Genetics, geographic variation, and community ecology

10.1 Animal athletics from a population perspective

A common quote is "variation is the spice of life," and that statement rings true in the context of the evolution of animal athletics. While much of this book has dealt with the notable differences in morphology and performance among disparate species, such variation typically begins as genetic mutations, which then translate into variation in phenotype or behavior within a population. This variation is then acted upon by natural selection. Those genetic variants that possess phenotypes that are favored by natural selection will survive and flourish, and this process ultimately leads to the evolution of complex structures that define species (Williams 1966; Kingsolver et al. 2001). Once a genetic mutation has spread within a population, differences can emerge that may lead to the nascent stages of speciation, that is, the formation of species. This process can be greatly expedited or facilitated if differences in habitat structure exist between geographic areas, such as between tropical and desert habitats (fig. 10.1). In some cases, these ecological transitions can be very distinct, resulting in several geographic "races," each of which is connected by gene flow, occurring in close proximity (Endler 1977; Avise 2000). This process of local adaptation means that, to understand how animal performance traits have evolved, we need to understand both the genetic basis of those traits and how different populations have specialized to different habitats. This approach takes us into the realm of population genetics and studies of geographic variation. Fortunately, the last decade has provided an emerging body of work on animal performance traits that examines both aspects.

In a related way, the study of how different species interact in a community is also informative for understanding the evolution of animal performance. Animals and plants exist within ecological communities in which ecological

Animal Athletes. Duncan J. Irschick and Timothy E. Higham. © Duncan J. Irschick and Timothy E. Higham 2016. Published in 2016 by Oxford University Press.

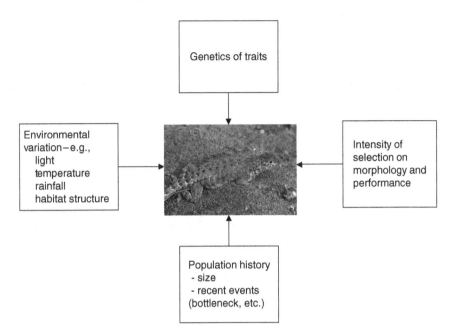

Fig. 10.1 A heuristic diagram showing different contributions to geographic variation in morphology and performance. Some of these influences include environmental differences among areas, the genetics of the traits involved, the population history, and also the nature of selection on traits.

and evolutionary forces can influence morphological shape and performance capacity. Within ecological communities, different species and individuals can impact one another directly (such as through predation) and indirectly (such as by competing for the same food resource). This area of research thus falls into the realm of community ecology, which is the study of how species within ecological communities utilize and partition resources and, ultimately, coexist. One of the primary goals of community ecology has been to generate workable models that explain how these various factors interact to produce functioning communities. One commonly held view is that certain resources are limited and that different individuals and species will compete for access to them (Cody and Diamond 1975). Competition for limited resources acts as a key constraint both on the total biomass that can be supported within a community and on how many different kinds of species can coexist. Examples of such limited resources include food (e.g., plants for herbivores, insects for herbivores), sunlight, the amount of available soil, and many others.

This chapter will consider both the genetic basis of performance traits and how performance traits vary among different populations. It will also address how variation in performance capacities enables different species within ecological communities to access various resources.

10.2 The genetic basis of animal athletics

A central pillar of evolutionary biology is that, for a trait to evolve, it must possess some genetic basis. In other words, some aspects of the trait must be passed on from parents to their offspring. Traits that have a strong genetic basis are more likely to quickly respond to evolutionary pressures than traits with a weak genetic basis. It is important to differentiate between selection on traits, and the evolutionary response to selection. Traits can experience strong selection but, in the absence of a genetic basis, evolution will not occur. On the other hand, even relatively weak selection on traits with some genetic basis can result in striking evolutionary changes over long periods. Because perform-ance traits emerge from underlying morphological, physiological, and behav-ioral traits, their genetic basis will be the primary limits for how performance traits evolve. The simplest way to understand how genes contribute to traits is through the concept of heritability. Values of heritability range from 0 (no gen-etic basis) to 1 (100% genetically controlled), so a heritability of 0.5 means that about 50% of the variation in the trait shows some genetic basis, whereas the other 50% is explained by other variables, such as the environment. The level of heritability may also offer some clues as to how natural selection either may have acted or is likely to act on the trait. It is generally thought that traits that have experienced intense selection are likely to have low heritabilities, as the additive variance has been depleted over time, although this view is debated. Paradoxically, a low heritability could also signify that the trait is less likely to evolve as a result of selection. However, there are many factors that influence heritability, making past or future inferences challenging. Many studies have examined the heritability of basic morphological traits, such as body weight, height, limb length, and so forth, in a wide variety of animals. In the majority of cases, morphological traits do display some level of heritability, although how much varies among species. Contrary to popular misconceptions, most morphological or performance traits are not controlled by a single gene but rather are controlled by multiple genes which interact with one another, a pat-tern known as pleiotropy.

While little data existed 30 years ago, there has been a steady increase in the number of studies examining the heritability of performance traits. In gen-eral, the heritability values for performance traits seem to fall within the same range as physiological or morphological variables but are somewhat higher than life-history variables. However, there is notable variation among traits and species, and this variation cannot be easily explained. For example, the average estimated heritability of maximum crawling speed for garter snakes, genus *Thamnophis*, is 0.58 (Garland 1988), but the estimated heritability for VO_{2max} in these same snakes is 0.88 (Garland and Bennett 1990). By compari-son, other biochemical traits in these species, such as the amount of liver citrate synthase, have heritabilities that are substantially lower, ranging from 0.01 to 0.58. In other species, some of these traits have very different genetic backgrounds. For example, the heritability of VO_{2max} in house mice barely differs from 0, meaning that few genes influence this measure in this species (Nespolo et al. 2003). In speckled wood butterflies, *Pararge aegeria*, the maximum ability to accelerate during takeoff, an important escape behavior for these animals, had a heritability of only 0.15, which was not significantly different from 0. One area in which a great deal is known about the heritability of performance traits is for horse racing, which employs selective breeding programs to obtain fast horses that will, in turn, earn their owners a lot of money—if they win! In an analysis of more than 400 racehorses, the speed to elicit 200 heartbeats/min (essentially a rough index of VO_{2max}) was found to have a heritability of 0.46 (Hinchcliff et al. 2013). In other research, heritabilities ranged from about 0.23 to 0.52 for a range of gait and jumping measures. These very high values, especially in contrast to some of the other values shown above, should be interpreted with caution, given the long history of domesticated breeding of racehorses. For example, the nonsignificant heritability of wild mice compared to a nearly 50% for domesticated racehorses suggests very different patterns of selection on this trait. In general, these examples show that many functional traits in a range of species can show quite high variabilities, although there is a great deal of variation among species and among traits.

In a few cases, there is evidence that performance traits can be influenced by genes whose expression can be modified in a relatively simple fashion. Invertebrates have been the best studied in this regard because one can easily breed individuals in laboratory settings. In moths and butterflies, expression of the gene *Troponin-t*, which plays a key role in the production of force and power in the muscles that control flight, is regulated both by the amount of food consumed by the insects when they are in the larval stage and by total adult body weight. This regulation occurs via a process called alternative splicing, through which different mRNA transcripts can be generated from a single DNA sequence; in this way, a single gene can give rise to multiple protein isoforms, each with different properties (fig. 10.2). Rick Marden and his colleagues (2008) showed that, when larvae were fed less, there was an increase in the amount of alternative splicing that took place; therefore, the relative abundance of the various Troponin-t (Tnt) protein isoforms was altered, thereby negatively influencing flight performance. In addition, by loading adult butterflies during flight, the researchers showed that the amount of alternative splicing that took place in response to the artificial loading was nearly identical to that induced by nutrition-generated increases in body weight. In other words, the Tnt isoform composition in lepidopteran striated muscle responds both to how much weight the animal must bear during flight and to the amount of food that the animal received during the larval stage. However, Tnt isoform composition is

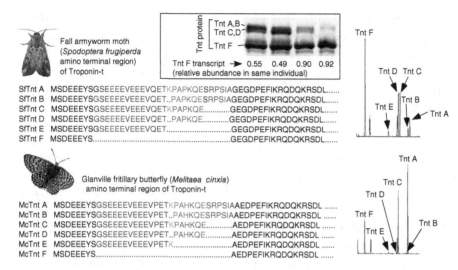

Fig. 10.2 Figure of fall armyworm moths (*Spodoptera frugiperda*) and Glanville fritillary (*Melitaea cinxia*) butterflies showing the variable N-terminal region of protein isoforms produced by alternative splicing of the *Tnt* gene. The *Tnt F* protein was the dominant form when *Tnt F* transcript was relatively abundant, but it was about half the protein when *Tnt F* represented only 50% of the transcript pool. Figure taken from Marden et al. (2008) with permission from the Company of Biologists Ltd.

also associated with activity level. These results suggest that isoform composition had an effect not only on flight performance per se but also on flight behavior and motivation. This study thus provides a compelling example of how a gene that affects muscle morphology (and consequently flight performance) can be affected by simple environmental factors. Further studies might show how variation in this gene affects different aspects of flight performance and how this might affect performance in nature.

Key genes seem to strongly influence metabolic rate and locomotion in butterflies. For many animals, the overall level of mobility may be important both for understanding how much energy is expended on a daily basis and for understanding how animals can disperse. In butterflies, the metabolic rate during flight positively impacts whether butterflies will disperse and is also correlated with the total amount of energy expended during flight. Some research indicates that the flight metabolic rate and frequency of variation in the phosphoglucose isomerase gene *Pgi* are linked. Those butterflies that possessed specific alleles of *Pgi* tended to have elevated flight metabolic rates. Because the rate of dispersal is closely tied to the population dynamics of this species, this work suggests that genetic variation of a performance trait may have significant impacts on ecological dynamics. Interestingly, butterflies typically cannot achieve flight until they reach a certain critical body temperature (which varies among species). Variation in *Pgi* is also linked to variation in the ability of some butterflies to effectively fly at low body temperatures; such an ability is potentially advantageous for females, as it allows them to fly and lay eggs when the temperatures are suboptimal. Specific genes also seem to play a role in performance traits in human athletic performance (discussed more in chapter 11). Some variants of mitochondrial DNA are linked to variation in VO_{2max}, and other variants are correlated with the ability to train aerobically (Dionne et al. 1991).

Correlations among traits can have a genetic basis and can affect how performance traits evolve. As explained in chapter 8, sexual signals may provide information on male quality, especially in the form of performance capacities that are important for fighting or mating. Male crickets, *Teleogryllus commodus*, vocalize to attract females, and there is variation among individuals in terms of how much time individuals are willing to spend calling. Simon Lailvaux and his colleagues (2010) have examined how female choice (i.e., attractiveness) interacts with calling effort and jumping performance, a key way for these animals to escape predators. It turns out that females do not prefer one kind of male but rather prefer males with different combinations of these traits, all of which show some level of genetic correlation (association) among them. This example shows how genetic linkage among multiple traits can lead to ambiguous outcomes of female choice and cautions against a simplistic view of the genetic basis of traits or of how they are related. Other studies with brown apple moths show that there are strong genetic correlations between some life-history traits and various metrics of flight capacity (Gu and Danthanarayana 1992a, b). These developmental traits include developmental time, and age at first reproduction, among others, and this intimate genetic interrelationship indicates that there may be significant trade-offs that can alter how selection on performance can occur.

10.3 Geographic variation in performance capacity

One of the most persistent problems facing any study of geographic variation is differentiating between adaptation via genetic change, and plastic change (Schlichting and Pigliucci 1998). For example, if one observes differences in maximum aerobic capacity between two populations of an animal species, but one population occurs at a high elevation, and the other occurs at a low elevation, then this physiological difference may be due to animals developing differently as a consequence of being raised at different partial pressures of oxygen. If offspring from these two populations were raised at a single low elevation site, this difference might then disappear. The common-garden method is perhaps the most common way to differentiate between plasticity and genetic causes of among-population variation. It involves taking samples from divergent populations and raising them in a common-laboratory setting. The basic idea behind this approach is that, if the documented differences are genetically based, then the differences will persist in the offspring from different populations. On the other hand, if the differences among populations are plastic, when offspring from divergent populations are placed in the same laboratory setting, the differences should disappear.

For populations that occur in different habitats, variation in predation pressure is perhaps the most driving force for morphology and performance. The role of predation for shaping the morphology and performance of wild animals has been investigated in some detail in guppies (Reznick and Endler 1982). Guppies on the island of Trinidad occur in various pools that vary in predation pressure. The differing levels of predation pressure come from different kinds and densities of predators. Experimental studies by David Reznick and his colleagues have shown that, in the presence of increased predation pressure, female guppies have evolved an increased allocation of resources toward reproduction. While reproduction is clearly important for these animals, they also operate under the constraint that the ability to swim rapidly is important for eluding aquatic predators, such as cichlid fish. These seemingly opposing factors result in a constrained form of evolution. Female guppies from high-predation localities reached higher accelerations and velocities and traveled longer distances within the swimming trials (Table 10.1; Ghalambor et al. 2004). However, velocity and distance traveled decline more rapidly over the course of pregnancy in these same females, thus reducing the magnitude of

Table 10.1. Burst-speed performance values for females from the Trinidadian guppy, *Poecilia reticulata*, comparing fish from low- and high-predation sites. Note that values of acceleration and velocity are higher in fish from high-predation populations. Taken from Ghalambor et al. (2004) with permission from University of Chicago Press.

Trait	Predation	
	Low	High
Body mass (mg)	447 (32)	491 (29)
Maximum acceleration (m/s/s)	53.25 (4.37)	67.35 (3.68)
Average acceleration (m/s/s)	33.37 (3.15)	41.54 (2.59)
Maximum velocity (m/s)	0.854 (0.03)	0.913 (0.03)
Average velocity (m/s)	0.594 (0.02)	0.615 (0.01)
Cumulative distance traveled (em)	1.32 (0.04)	1.37 (0.03)
Total turning angle (deg/s)	11,000 (558)	11,527 (520)
Average rotational velocity (deg)	101 (6.5)	103 (6.1)

Note: Shown are least square means (+ SE) from general linear models, except for maximum acceleration and body mass.

divergence in swimming performance between high- and low-predation communities. Hence, adaptive evolution in swimming performance in Trinidad guppies appears to be in part constrained by selection for enhanced reproductive allocation.

The need to move effectively in different habitats can generate significant among-population variation in overall body shape. Common-garden studies in threespine stickleback (*Gasterosteus aculeatus*) show that there are both genetic and plastic components to swimming performance and that both components influence the fish's ability to specialize for different habitats. In general, stickleback fish that reside in lakes have large caudal fins, shallow bodies, and relatively narrow pelvises. Lake stickleback are also superior performers in terms of sustained and burst swimming. By comparison, stickleback that reside in inlet areas may be less speedy, but they are more maneuverable. Interestingly, the same morphological traits do not influence performance traits in the same way across these two populations. The fact that morphology-performance relationships can diverge across even relatively short geographical expanses suggests that the genetic architecture of morphological and performance traits can be altered relatively quickly. When populations of the same species differ as to whether they face extreme physical challenges, selection for performance capacities can cause among-population variation (Garland and Adolph 1991). Anyone who has watched fish migrating cannot help but sympathize at their remarkable plight as they swim upstream against both rapids and hungry predators. It turns out that, within some species, there is variation among populations in relation to the length of migration. Anadromous cisco fish, *Coregonus artedii*, that must swim a longer distance upstream during migration have higher aerobic capacities than those that must swim a shorter distance, and this ability appears to be due to physiological differences in their musculature (Couture and Guderley 1990).

Survival in different environments often presents trade-offs that can result in differences among populations. Specialization for different thermal environments often creates some of the most notable differences among taxa. For example, silversides are small ubiquitous fish that have a vast geographic range from warm southern waters to much cooler northern waters. Trade-offs occur in northern and southern populations of silversides in terms of their growth rates and how growth influences locomotor performance. In general, cold northern habitats offer a short growing season; consequently, development must be rapid in order for juveniles to successfully overwinter. Thus, Atlantic silversides that inhabit cool, northern waters have fast growth rates, whereas Atlantic silversides that live in warm, southern waters have slow growth rates (Conover and Present 1990), and the difference between the growth rates of these two populations can be up to threefold. Interestingly, rapid growth may impose costs on locomotor performance, as northern silverside populations have poorer aerobic and burst abilities compared to southern silverside populations. Experimental studies show that the inducement of rapid growth directly results in reduced locomotor performance and a reduced ability to elude fish predators. The reason why growth has a negative impact on locomotion in these fish is not fully understood but may be that rapid growth draws away resources from tissues and structures that drive locomotion (e.g., some muscle groups, and fins).

In addition, specialization to different thermal environments may impose trade-offs in the ability of populations from different areas to perform well across a variety of thermal environments. For example, Australian striped frogs, *Limnodynastes peronii*, occur across a wide range of thermal habitats, from the hot tropical north to the somewhat cooler south. Since this species is an ectotherm and occurs across this wide expanse, we can predict that populations that live in cool southern environments should be superior jumpers at cool temperatures but that populations that live in hot northern environments should be superior at warm temperatures. This prediction is largely upheld, but there are some variables that are hardly affected, such as the maximum force and acceleration of jumping, while others, such as maximum jump distance and takeoff velocity, differ far more among populations. Moreover, when this experiment was repeated with laboratory-raised young individuals from these populations, these patterns remained consistent; this result suggests that there is some genetic component to this among-population thermal sensitivity.

10.4 The role of community structure in molding animal athletics

In the context of community ecology, early work (done 30–40 years ago) focused on resource limitation as the primary factor influencing community structure. These same studies often focused on morphological differences among species and assumed that morphological variation should reflect differences in resource use. A direct relationship between morphology and function may be true for some species, but it is not universal. Organisms may differ morphologically yet be functionally equivalent, and this phenomenon is often referred to as many-to-one mapping of form to function (Wainwright et al. 2005). Such mapping is common in a wide range of taxa, including lizards (Toro et al. 2004) and fishes. Additionally, species may be morphologically similar yet functionally divergent. There is an emerging awareness that direct measurements of body form and performance capacity may be vital to helping us understand resource use within communities. As a point of definition, an ecological community consists of many different kinds of species, such as insects, vertebrates, plants, and other organisms, that occur in a single location and that clearly interact. In practice, because of this overwhelming complexity, many researchers focus on a single animal type, such as all the species of fishes within a community.

Anyone who has walked outside in the tropics at dusk has likely experienced the rapid rush of air from the fluttering wings of bats zipping by one's ears. Bats are among the most important members of many ecological communities, as they play a key role in regulating invertebrate populations and pollinating specialized plant species. However, bats not only consume insects but have diversified to consume a wide array of food, including fruit, frogs, mammals, reptiles, nectar, and blood, among others. While bats vary noticeably in overall body size and shape, it is the variation in the shape of their heads that stands out, with bat heads varying in the length of the snout, teeth, and other features (fig. 10.3). The fact that bats consume such a wide array of food types

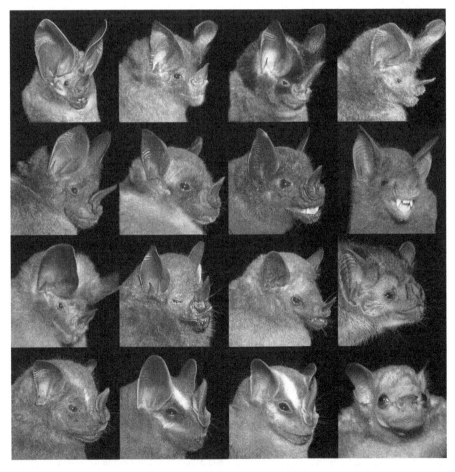

Fig. 10.3 A figure showing the diversity of phyllostomid bats, particularly in relation to their head morphology, which is closely tied to their feeding habits. Image by Sharlene Santana.

suggests that there should also be variation in biting performance. However, in a survey of 20 bat species in a tropical African savannah community, biting force did not differ according to diet, at least among the major groups of bats (insectivores, frugivores, and omnivores; Aguirre et al. 2002; see fig. 10.4). This result suggests that bats can specialize on very different prey items while still using about the same level of biting force. However, once one compares bats with highly specialized diets (nectar-eating, fish-eating, and blood-eating bats) to these other species, one finds that specialization to these unusual food types does not require strong bites; therefore specialization in this community largely consists of a single mode (high bite performance; see fig. 10.4), with any reduction in biting performance occurring for a simple reason: bats that consume fish, nectar, or blood apparently don't require strong bites!

A second example comes from one of our own research projects (D. Irschick). Much like the bat example above, lizards coexist sympatrically in desert communities in the Southwestern United States. Unlike the bat example above, however, these lizards overlap considerably in their prey choice—namely, small insects. In a study site in a mid-elevation riparian zone in northern Arizona, 10 lizard species that vary in size from a few grams to over 100 grams in body mass can be found at the same site (Meyers and Irschick 2015). The existence of this group of lizards poses a fascinating question: given that all of these lizards capture and crush insect prey with their jaws, does higher biting

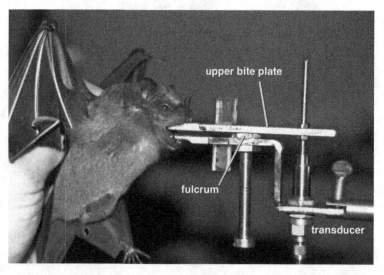

Fig. 10.4 Device for quantifying bite force in bats. Animals are suspended by their wings in order to provide free access to the bite plates. The upper bite plate rotates across the fulcrum as a result of biting and consequently pulls the transducer. Image taken from Aguirre et al. (2002) with permission from the Royal Society Publishing Group.

performance open the gateway to consuming a wider range of prey? By analyzing both how hard these lizards can bite and how much force is required
to crush the available insect prey, it's possible to answer this question. Much
of the variation in bite force is related to body size, as larger lizard species
in this community also bite harder because their heads are larger. However,
there is also variation among species in how hard they bite once the effects of
size are accounted for, as 6 of the 10 species can crush and ingest over 95% of
the prey that is available to them. The other 4 species can still consume about
75% of the prey that is available to them. This means that larger lizards in this
community have access to a much wider food base than smaller lizards do
and that, rather than thinking of these lizards as occupying separate niches
in terms of diet, they can be better thought as overlapping "Russian dolls"
in which larger species are the larger "doll," and the smaller species are the
smaller "dolls."

In both of these above examples, bite force is one of many ways to define
how these species access resources and coexist, but these analyses reveal two
important lessons. First, access to resources is often not always dictated by an
amplification of functional abilities but rather can occur through a diminishment of these abilities. Evolution only works toward what is "good enough,"
and excess abilities are rarely favored by natural selection. Second, changes
in body size can dramatically alter the ability of species to access certain resources, above and beyond size-specific changes in morphology or behavior.

Islands offer an outstanding opportunity to evaluate how species within
a community have diverged and how a functional approach can inform us
regarding the nature of diversification. One well-known example is the Darwin's finches, genus *Geospiza*, of the Galapagos Islands in Ecuador. Darwin's
finches are small ground finches native to the Galapagos and are partially responsible for inspiring Darwin's theory of evolution via natural selection. Different species occur on different islands within the Galapagos archipelago, but
long-term research by Peter and Rosemary Grant of Princeton University has
focused on several coexisting species on an island called Isla Daphne Major.
For the most part, these birds survive on a diet of hard seeds that they crush
with their beaks. Some finch species have deep and powerful beaks that are
used to crush hard prey, whereas others species have more gracile beaks that
are better suited for consuming smaller seeds. This relationship between beak
shape and resource use among species likely occurs because of the biomechanical relationship between beak morphology, bite force, and the ability to
crush seeds.

If the environment were static, that would be the end of the story; but the
Galapagos Islands undergo annual dry and wet cycles because of a cyclical
weather event called El Niño. This weather pattern shifts the island from
being densely vegetated, with abundant seed types and sizes, to extremely

xeric, with relatively few seeds and seed types. Behavioral research has shown that Galapagos finches exhibit complex behaviors and choices in terms of the seeds and food types. Moreover, different finch species use different foraging behaviors for different kinds of seeds. During flush periods, finches can choose between hard and soft seeds whereas, in dry periods, the majority of seeds are hard and difficult to crack. This ecological dynamic poses a functional challenge: individuals or species that can crush hard seeds (i.e., with high seed-crushing ability) can access food at all times of year, whereas individuals or species with less beak strength are more limited in terms of food availability. In short, performance capacities (bite force) interact with resource use to influence species-level differences, as well as variation in fitness among individuals within species.

Not surprisingly, another example of a relatively simple community structure comes from islands as well. The *Anolis* lizards of the Caribbean are a diverse and speciose group, with over 400 species in Central America, South America, and the Bahamas, but it is the anoles of the Caribbean that have received the most attention (Losos 2009). There are several hundred species of *Anolis* lizards, of which a large portion live in the Caribbean, and w species differ notably in their limb and body proportions, as well as in behavior and habitat use. Larger Caribbean islands possess assemblages of anole lizard species of varying numbers (e.g., over 50 in Cuba and Hispaniola, and 7 in Jamaica), and each island has several "ecomorphs," which are species that occupy unique habitats such as tree crowns or narrow twigs and have corresponding differences in limb, body, and tail dimensions, as well as performance capacities. There is a direct relationship between average habitat use and locomotor performance in these species (Losos and Sinervo 1989; Irschick and Losos 1999). Anole ecomorphs that occur on broad surfaces such as tree-trunks (the "trunk-ground anoles") tend to have long hindlimbs and are fast runners, whereas those that occupy narrow surfaces (the "twig anoles") tend to have short hindlimbs and move slowly. This differential importance of locomotor speed is closely tied to the ecological strategies of the anole ecomorphs. The twig anoles rely on crypsis, not speed, to elude predators and capture prey, whereas trunk-ground anoles use speed for both behaviors, as well as for others. In turn, the different limb lengths (long vs. short) enable these animals to move effectively on either broad (trunk-ground) or narrow (twig) perches (fig. 10.5). This across-island matching of morphology, habitat use, and performance capacities occurs with great consistency (Irschick and Losos 1998) and demonstrates a clear example of a direct relation between resource use (perch sizes) and performance.

Whereas food resources and the use of perches influence morphology in the above examples, other environmental variables exert a more physical and demanding impact on species. For coral reef fishes in the Great Barrier Reef, the physical demands of moving in turbulent water appear to have shaped, in an evolutionary sense, the shape and function of their fins. Variation among coral reef habitats in both fin shape and swimming performance was compared against

Fig. 10.5 An example of how variation in habitat structure affects animal locomotor performance. *Left*, Images of three lizard species that occupy different substrates, notably different surface diameters (top to bottom): *Anolis sagrei, Anolis smaragdinus,* and *Anolis angusticeps. Anolis sagrei,* a trunk-ground anole, prefers broad substrates, whereas the other two species, especially *Anolis angusticeps,* a twig anole, prefers narrower substrates. *Right*, The relationship between substrate diameter (x-axis) and maximum sprint speed as measured under laboratory conditions in several *Anolis* lizard species. The degree of change (sprint sensitivity) in sprint speed to substrate diameter is related to the ecology and morphology of the species. Species with long hindlimbs, such as *A. lineatopus* or *A. sagrei,* suffer a substantial decline in sprint speed as dowel diameter becomes narrower. On the other hand, species with short hindlimbs, such as *A. valencienni,* run equally well on substrates of differing diameters. Taken from Irschick and Losos (1999) with permission from the University of Chicago Press. Images of lizards are from Duncan J. Irschick.

values of water motion (from waves) on the Great Barrier Reef (Fulton et al. 2005). How fast a fish choses to swim in the field (both field and laboratory measures of speed were examined) is correlated with the degree of water motion derived from wave action (fig. 10.6). This means that coral reef fishes have evolved effective ways of moving in turbulent habitats by evolving morphological structures that facilitate locomotion in those environments. Although the velocity and other characteristics of flow are not "resources" in the same sense as seeds in birds, they still represent powerful force shaping these aquatic communities.

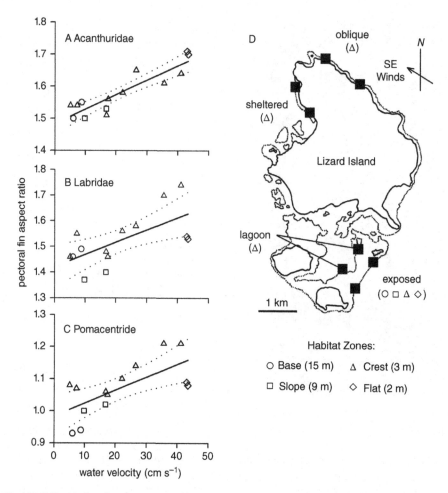

Fig. 10.6 Correlations between pectoral fin aspect ratio and net water velocity for three families of coral reef fishes. A, Acanthuridae. B, Labridae. C, Pomacentridae. Dotted lines indicate the 95% confidence limits. D, The locations of the habitats where fishes were sampled, including four different exposures around Lizard Island. Figure taken from Fulton et al. (2005) with permission from the Royal Society Publishing Group.

10.5 Many-to-one mapping, and communities

Despite the above examples of obvious and straightforward resource partitioning, recent work has shown that phenotypic and performance variation may not be related to resource use in a straightforward fashion. Indeed, it is unlikely that one-to-one mapping is the norm for most communities or evolutionary radiations. Many phenotypic traits have the capacity to exhibit multiple functions and so can perform several different tasks (Wainwright et al. 2005).

Coral reefs represent some of the most diverse communities in the world, and the fishes that live in these reefs provide a test case for examining linkages between performance capacities, resource use, and morphology. At first glance, fish jaws present great potential for multifunctionality: they are composed of a complex series of levers (jaw bones, muscles, and soft tissue) that can be moved and used in a variety of ways (e.g., by projecting certain mouthparts, or for suction). Fish jaws are complex structures but can be simplified into several basic components. The lower and upper jaws form a lever system for which we can calculate simple measures of mechanical advantage. Jaws that have a high mechanical advantage will have a superior capacity to crush prey that is larger, harder, or both, whereas fish jaws with a lower mechanical advantage should not be able to crush such prey. According to this simple hypothesis, one should observe a direct correlation between the size and shape of fish jaws and the prey the fishes eat; however, analyses both of large groups of fish radiations and of communities suggest otherwise. Fishes feed in myriad ways, ranging from suction feeding (in which no real biting or teeth are required) to durophagy (the crushing of hard prey, such as snails; typically requires a robust jaw and teeth). Research by Peter Wainwright and his colleagues (2005) has shown that the jaws of fishes from large and complex coral reef communities or from larger fish evolutionary radiations seem to exhibit both "many-to-one" and "one-to-many" mapping of fish jaw lever systems (a proxy for functional data on fish jaws) and the kind of prey they consume (e.g., suction feeding, durophagy; fig. 10.7).

The "generality" of animal jaws for diet, as well as the lack of specialization among closely related species, has been demonstrated in other species. Another example comes from the ability to accelerate and move at high speeds in anoles (Vanhooydonck et al. 2006). The ability of an animal to accelerate over short distances is often one of the most important performance attributes for evading a hungry predator. Because high accelerations require high amounts of mechanical power, large muscle volume per unit of body mass is the primary explanatory factor for enhanced acceleration among species of anole, with higher accelerators having more muscular legs than lower accelerators. By contrast, the ability to reach high top speeds is driven largely in lizards by the length of the hindlimb. Therefore, lizards with long, muscular limbs can achieve both ends—they can obtain high accelerations and high top speeds (fig. 10.8). For jumping in anoles it appears that there are three ways to be a good jumper (i.e., to be able to make long jumps): evolve long limbs, evolve muscular limbs, or both (Toro et al. 2004). As noted above, studies with bats also show that similar levels of bite performance are found in bats that consume a wide range of prey (i.e., omnivores, frugivores, and insectivores), although when bats consume more unusual food, such as blood, there seem to be changes in overall feeding and bite performance (Aguirre et al. 2002).

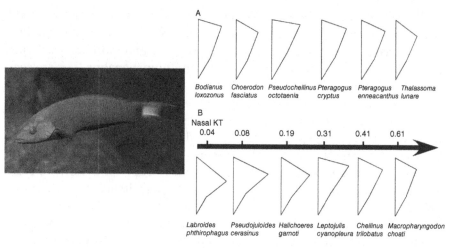

Fig. 10.7 An example of how similar mechanical properties can be obtained with different shapes. *Left*, Image of the labrid fish *Thalassoma lunare*. *Right* (*A–B*), Maxillary kinematic transmission coefficient (maxillary KT) and nasal kinematic transmission coefficient (nasal KT) in a range of labrid fish species. *A*, Six species of labrid fishes that share the same values for maxillary KT and nasal KT in the oral jaw four-bar linkage but achieve these values by using different shapes. *B*, Six species of labrid fishes that exhibit the same value for maxillary KT but for which values of nasal KT differ. Figure taken from Wainwright et al. (2005) with permission from Oxford University Press. Image of *Thalassoma lunare* taken from Wikimedia Commons.

Fig. 10.8 *Left*, Image of an *Anolis* lizard. *Right*, Plot showing the positive evolutionary relationship between maximum sprint speed and maximum acceleration among a group of *Anolis* lizard species. As noted in the text, lizards can maximize both kinds of performance by evolving muscular limbs (for increased power, and thus increased acceleration), long limbs (for increasing stride length, and hence maximum speed), or both. Taken from Vanhooydonck et al. (2006) with permission from Wiley-Blackwell Press. Image of lizard is from Duncan J. Irschick.

This result could be viewed as a form of many-to-one mapping because having similar levels of performance doesn't restrict food resource acquisition, at least for a range of food items. A final example may come from principles of vertebrate muscle, as organisms can evolve a given volume of muscle in a variety of different ways by combing elements such as the length and orientation of the fibers, as well as other factors (Powell et al. 1984; Wainwright et al. 2005). Many other examples likely exist and remain to be discovered and elaborated upon.

References

Aguirre LF, Herrel A, Van Damme R, Matthysen E. 2002. Ecomorphological analysis of trophic niche partitioning in a tropical savannah bat community. Proceedings of the Royal Society B 269:1271–1278.

Avise JC. 2000. Phylogeography: The History and Formation of Species. Harvard University Press, Cambridge, MA.

Cody ML, Diamond JM. 1975. Ecology and Evolution of Communities. Harvard University Press, Cambridge, MA.

Conover DO, Present TMC. 1990. Countergradient variation in growth rate: Compensation for length of the growing season among Atlantic silversides from different latitudes. Oecologia 83:316–324.

Couture P, Guderley H. 1990. Metabolic organization in swimming muscle of anadromous coregonines from James and Hudson bays. Canadian Journal of Zoology 68:1552–1558.

Dionne FT, Turcotte L, Thibault MC, Boulay MR, Skinner JS, Bouchard C. 1991. Mitochondrial DNA sequence polymorphism, VO2max, and response to endurance training. Medicine and Science in Sports and Exercise 23:177–185.

Endler JA. 1977. Geographic Variation, Speciation and Clines. Princeton University Press, Princeton, NJ.

Fulton CJ, Bellwood DR, Wainwright PC. 2005. Wave energy and swimming performance shape coral reef fish assemblages. Proceedings of the Royal Society B 272:827–832.

Garland T Jr. 1988. Genetic basis of activity metabolism. I. Inheritance of speed, stamina, and antipredator displays in the garter snake *Thamnophis sirtalis*. Evolution 42:335–350.

Garland T Jr., Adolph SC. 1991. Physiological differentiation of vertebrate populations. Annual Review of Ecology and Systematics 22:193–228.

Garland T Jr., Bennett AF. 1990. Quantitative genetics of maximal oxygen consumption in a garter snake. American Journal of Physiology 259:R986–R992.

Ghalambor CK, Reznick DN, Walker JA. 2004. Constraints on adaptive evolution: The functional trade-off between reproduction and fast-start swimming performance in the Trinidadian guppy (*Poecilia reticulata*). American Naturalist 164:38–50.

Gu H, Danthanarayana W. 1992a. Quantitative genetic analysis of dispersal in *Epiphyas postvittana*. I. Genetic variation in flight capacity. Heredity 68:53–60.

Gu H, Danthanarayana W. 1992b. Quantitative genetic analysis of dispersal in *Epiphyas postvittana*. II. Genetic covariations between flight capacity and life-history traits. Heredity 68:61–69.

Hinchcliff KW, Kaneps AJ, Geor RJ. 2013. Equine Sports Medicine and Surgery. Elsevier Health Sciences, Oxford.

Irschick DJ, Losos JB. 1998. A comparative analysis of the ecological significance of maximal locomotor performance in Caribbean *Anolis* lizards. Evolution 52:219–226.

Irschick DJ, Losos JB. 1999. Do lizards avoid habitats in which their performance is submaximal? The relationship between sprinting capabilities and structural habitat use in Caribbean anoles. American Naturalist 154:293–305.

Kingsolver JG, Hoekstra HE, Hoekstra JM, Berrigan D, Vignieri SN, Hill CE, Hoang A, Gibert P, Beerli P. 2001. The strength of phenotypic selection in natural populations. American Naturalist 157:245–261.

Lailvaux SP, Hall MD, Brooks RC. 2010. Performance is not proxy for genetic quality: Trade-offs between locomotion, attractiveness, and life history in crickets. Ecology 91:1530–1537.

Losos JB. 2009. Lizards in an Evolutionary Tree: Ecology and Adaptive Radiation of Anoles. University of California Press, Berkeley. CA.

Losos JB, Sinervo B. 1989. The effects of morphology and perch diameter on sprint performance of *Anolis* lizards. Journal of Experimental Biology 145:23–30.

Marden JH, Fescemyer HW, Saastamoinen M, McFarland SP, Vera JC, Frilander MJ, Hanski I. 2008. Weight and nutrition affect pre-mRNA splicing of a muscle gene associated with performance, energetics and life history. Journal of Experimental Biology 211:3653–3660.

Meyers JJ, Irschick DJ. 2015. Does whole-organism performance constrain resource use? A community test with desert lizards. Biological Journal of the Linnean Society. doi: 10.1111/bij.12537.

Nespolo RF, Bacigalupe LD, Bozinovic F. 2003. Heritability of energetics in a wild mammal, the leaf-eared mouse (*Phyllotis darwini*). Evolution 57:1679–1688.

Powell P, Roy RR, Kanim P, Bello MA, Edgerton V. 1984. Predictability of muscle tension from architectural determinations in guinea pig hindlimbs. Journal of Applied Physiology 57:1715–1721.

Reznick D, Endler JA. 1982. The impact of predation on life history evolution in Trinidadian guppies (*Poecilia reticulata*). Evolution 36:160–177.

Schlichting CD, Pigliucci M. 1998. Phenotypic Evolution: A Reaction Norm Perspective. Sinauer Associates, Sunderland, MA.

Toro E, Herrel A, Irschick DJ. 2004. The evolution of jumping performance in Caribbean *Anolis* lizards: Solutions to biomechanical trade-offs. American Naturalist 163:844–856.

Vanhooydonck B, Herrel A, Van Damme R, Irschick DJ. 2006. The quick and the fast: The evolution of acceleration capacity in *Anolis* lizards. Evolution 60:2137–2147.

Wainwright PC, Alfaro ME, Bolnick DI, Hulsey CD. 2005. Many-to-one mapping of form to function: A general principle in organismal design? Integrative and Comparative Biology 45:256–262.

Williams GC. 1966. Adaptation and Natural Selection. Princeton University Press, Princeton, NJ.

11 | Human athletics
A link to nonhuman animals?

11.1 Humans versus other animals

So far, we have focused on nonhuman animals, except to point out how much of what other animals can do we cannot! As scientists, we are often asked to justify our research, and the expected (but perhaps wrong) answer from most people is that, by studying nonhuman animals, we can gain some insight into humans. Surprisingly, a large portion of the time, the behaviors and abilities of most nonhuman animals are so distinct from humans that this is often an overstatement. Nevertheless, it is tempting to examine the differences between humans and other animals and ask whether any of the lessons learned from studies of nonhuman animal performance can be applied to humans (and vice versa). Fortunately, because of organized sporting events, there is a large body of data on human performance during many tasks, such as running, weight lifting, and so forth. There are some notable differences between human performance capacities and those of other animals. In nonhuman animals, these capacities have evolved over millions of years for one purpose: survival of the species. For most wild animals, this dynamic remains, although in some cases it has been ameliorated because of habitat disturbance, which can lead to the loss of key predators, among other factors. While performance abilities in humans have also evolved across our evolutionary history, one could argue that their original utility has faded since the advent of more complex civilizations. Thus, the evolutionary pressures that once shaped our athletic abilities have largely faded for most (though not all) humans. Most people in New York City do not have to worry about fleeing from animal predators!

Nevertheless, despite this shift in evolutionary pressures, humans remain quite impressive relative to other animals at certain tasks. Anyone who has

Animal Athletes. Duncan J. Irschick and Timothy E. Higham. © Duncan J. Irschick and Timothy E. Higham 2016. Published in 2016 by Oxford University Press.

watched the New York City Marathon or the Boston Marathon cannot help but be impressed with the remarkable running abilities of these highly trained athletes. In 2014, the Kenyan runner Dennis Kimetto broke the marathon world record with a time of 2 h, 2 min, and 57 s, which is average rate of approximately 4 min and 44 s per mile! The aerobic capacity of humans is not bad at all, even relative to wild animals with good aerobic capacities, such as dogs and horses. Indeed, humans possess many features that would seem to promote good endurance, such as a sophisticated system of capillaries that supply blood flow to peripheral tissues, an efficient heart that can pump large amounts of blood, and large numbers of mitochondria that fuel aerobic metabolism. The ability to run long distances may have arisen from the early conditions that shaped human culture and evolution. Anthropologists and paleontologists believe that early humans evolved as hunter-gatherers in large open areas of Africa, such as the Great Rift Valley. Because human groups were likely widely dispersed, as were potential prey, hunter-gatherers likely moved long distances to find suitable resources for survival. It is important to keep in mind that human survival in the early phase of human evolution was likely more precarious than in current times, at least compared to most modern societies. Individual variation in the ability to run long distances or to run quickly may have meant the difference between finding dinner and becoming dinner. While the reasons for our impressive aerobic performance may never be fully understood, it is clear that our bodies and abilities have also been shaped by evolution, in much the same manner as in a shark or a bird. Thus, we can ask whether the factors that dictate individual variation in performance capacity in humans follow the same patterns as in animals.

11.2 Performance specialists versus generalists

The human Olympics is a wonderful showcase of extreme specialization. Athletes train for years or even decades to perform an often highly regimented task that can sometimes last only a few seconds. This extreme specialization is enhanced by mental and physical training that often focuses on developing one part of the body. While these athletes can be highly effective at these tasks, does this extreme specialization result in compromised abilities for other tasks? As noted in chapter 7, a trade-off between the ability to perform different tasks well versus effectiveness at a single task seems to be a ubiquitous feature of evolution, though it is by no means a strict rule. Decathletes would seem to be a good group in which to address this question, as they compete in the same events as more specialized athletes, such as the 100 m dash, but must also perform well at several other tasks, such as middle-distance running.

Fig. 11.1 Images of athletes competing at the kinds of events used in a decathlon. These athletes compete in a range of tasks, such as the 100 m dash, the 1,500 m run, and the discus throw, activities that might seem to conflict with one another because of their different demands for speed versus endurance. Research by Raoul Van Damme and colleagues (2002) shows that the ability to specialize at one task seems to come at the expense of performing well at all of them. Note that the images are for various events, and the athletes may not be decathletes. Images from Wikimedia Commons.

These different tasks would seem to favor different kinds of phenotypes; so, are champion decathletes universally good at every event, or do they sacrifice performance at one task to perform well at another? In an analysis of athletic performance across ten different events, it appears that the ability to perform well overall (i.e., be generally good at all events) and the ability to perform well at any one event oppose one another (fig. 11.1; Van Damme et al. 2002). In other words, it appears that decathletes cannot excel at one event without sacrificing performance at other events, thus presenting a clear example of the "jack-of-all-trades-master-of-none" concept. Further, this finding suggests that decathletes may have to make careful strategic choices about how they will allocate their time to training for various tasks.

11.3 Training impacts

Humans train to perform better at athletic events. Professional athletes will spend tens of thousands of hours spread over decades to build muscle, endurance, or other factors. It is well established that exercise in humans dramatically improves athletic performance. Aerobic training results in increased numbers of mitochondria, a richer network of capillaries that deliver oxygen to various tissues, an improved ability to pump blood from the heart, and other benefits (Wang et al. 2004; Wilson 2013). Intense and prolonged endurance training can result in about a 30% increase in the density of capillaries and about a 40% elevation of mitochondrial volume, relative the levels seen at baseline (untrained) conditions (Hoppeler et al. 1985; Hoppeler and Fluck 2002). Training also has a great impact on strength. Weight training increases the average diameters of muscles in the body part being exercised, resulting in a large increase in muscle size after a period of hard training. For example, a low-repetition training regime can result in a 20% increase in the cross-sectional area of type IIB muscle fibers (Campos et al. 2002). However, variation among individuals in their response to training is well known; many of you likely know some people who can train hard yet show little improvement, while others seem to train hardly at all yet improve by leaps and bounds. It turns out that this scenario is more than urban legend— it has some merit. There is significant variation among individual humans in the ability of their bodies to respond to training, and there is some evidence that this may be due to genetic factors, such as the *ACE* gene. A particular allele of this gene is often found at high levels in elite athletes who perform extreme aerobic feats. The presence of this allele also seems to facilitate the impacts of training: individuals that carry this allele not only seem to possess enhanced aerobic capacities but also seem to improve more with training compared with individuals who do not carry it. While the function of *ACE* and its various alleles remains an active area of research, this gene appears to play a role in regulating the ability to contract muscles. Furthermore, it confirms that life is not fair!

Whether training also improves performance capacities in wild or domestic animals is less understood and remains an active area of study. In mammals, especially domestic animals, training seems to show a similar effect as for humans. For instance, when cats and rodents have been trained to jump to obtain food, or to pull on heavy baskets of food for several weeks, several muscle groups appear to increase in size. This effect appears to occur both in slow muscles and in fast muscles, such as the soleus muscle, which can enlarge by as much as 70% (Goldspink 1991). It appears that, as in humans, this enlargement of the muscle is mostly due to an increase in the size of individual fibers. The reasons for this similar response between humans and domestic or wild mammals may be due to similarities in underlying physiology and anatomy, as well as some similarities in lifestyle. Training seems to also improve aerobic or other kinds of swimming performance in various species of fishes.

By contrast, in squamate reptiles, training has mixed effects. In fence lizards, *Sceloporus occidentalis*, six-lined racerunners, *Aspidoscelis sexlineata*, and an agamid lizard, *Amphibolorus nuchalis*, aerobic training had little impact on aerobic capacity, relative to controls that simply rested in their cages (Gleeson 1979; Garland et al. 1987; O'Connor et al. 2011). Further, sprinting training actually caused a worsening in sprint speed in agamid lizards. Interestingly, intensive training appears to actually result in physical impairments in limbs and other body parts in many species; this result suggests that training is both unnatural and harmful in some cases. However, in a recent study with *Anolis* lizards (Husak et al. 2015), aerobic training appeared to elicit many of the same performance and physiological benefits as demonstrated in humans, such as

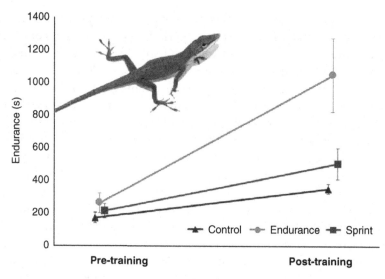

Fig. 11.2 The effects of sprint and endurance training on endurance capacity in green anoles, *Anolis carolinensis*. Those animals trained using an endurance protocol exhibited elevated endurance capacity. Values are means plus or minus the standard errors of the means. Redrawn from Husak et al. (2015) with permission from the Company of Biologists Ltd.

improved aerobic performance (fig. 11.2), an increase in hematocrit levels, and an increase in the size of fast glycolytic muscle fibers. Differences in lifestyle and physiology may explain why some lizards don't respond to exercise in the same manner as humans do. Lizards are typically ambush predators that rely on short bursts of speed to capture prey and elude predators. Thus, their aerobic capacities are poor, and they generally rely on anaerobic metabolism to fuel most movement. Nonetheless, the fact that at least some lizards can improve with exercise suggests that this description of their physiology is not universal. In addition, the stress response incurred in animals as a result of training is a poorly understood factor. Forcing a wild animal to run on a treadmill will induce high levels of stress hormones, which can compromise many aspects of the immune system and which might impair physiological improvement. While we might enjoy a run on treadmill, the same cannot be said for most wild animals! As reptiles exhibit a very strong stress response, one explanation for the lack of response in some lizards might be the harmful effects of stress hormones on aerobic physiology in some species.

One of the most interesting aspects of training for humans concerns decisions about how "hard" athletes should train and under which regimen of rest. Anyone who has watched an animal move likely has noticed the "intermittent" nature of their movements. Many lizards, mammals, fishes, and other animals tend to alternate bursts of higher speed movement with periods of rest. Even when eluding a predator, animals often apply this intermittent approach by making brief pauses between short bursts until the threat has subsided. Intensive studies by Randi Weinstein show how ghost crabs inter-mix high-intensity movements with periods of rest to prevent fatigue, a be-havior that is especially noticeable when they are stressed (Weinstein 1995). Physiological studies reveal that energy savings are the likely reason for this behavior. Quick bursts of movement typically occur above the maximum aer-obic speed for an animal, that is, the speed at which the animal uses both oxidative metabolism and nonoxidative pathways simultaneously and there-fore unsustainably, because of the rapid onset of fatigue. Anaerobic movement also tends to degrade levels of creatine phosphate; however, brief intervals in-between bouts of high-activity movement can enable the recovery of this substance. In short, some animals can increase the total distance moved if they intersperse high-intensity periods with brief periods of rest, rather than mov-ing at a continuous level of high intensity (fig. 11.3). Given that many human sports employ a similar mix of high-intensity movement with pauses of vary-ing duration (racquetball and soccer are two good examples), are the same benefits observed? A large body of research seems to indicate the answer is yes. Observations on soccer players shows that they will average about 11 km

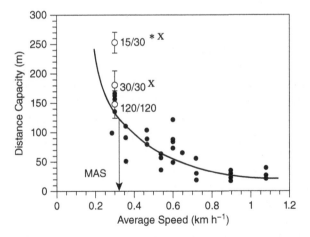

Fig. 11.3 A figure showing average speed versus distance capacity for intermittent versus continu-ous exercise for the gecko *Teratoscincus przewalskii*. The filled symbols and solid curve represent the distance capacity when these geckos are moving continuously. The open symbols are the distance capacity for intermittent locomotion. Values are means plus or minus the standard errors of the means. The numbers adjacent to open symbols indicate the intermittent exercise proced-ure (see Weinstein and Full 1999 for more details). The asterisks designate significant difference from continuous locomotion for the same mean speed. The crosses designate significant difference from continuous locomotion regarding the same absolute speed. Redrawn from Weinstein and Full (1999) with permission from the University of Chicago Press.

during a match, but players will typically only run at high intensity (usually lasting only a few seconds) for about 8% of their playing time. Players then intermix "rest" periods (walking or standing) of several seconds before moving again at a higher intensity (Bangsbo et al. 1991; Weinstein 2001). Indeed, interval training, which intermixes high-intensity activity with brief periods of rest, enables athletes to perform a higher total volume of "work" than does training continuously at a high intensity.

11.4 Steroids and performance

Anyone paying attention to the news media these days quickly realizes the pervasive problem of steroid abuse in human athletes. Any casual sports fan knows that steroids and other performance-enhancing drugs are rampant in professional or even amateur sports, especially in sports that require strength. The notable increase in the total numbers of home runs hit during the "steroid era" in baseball (about 1990 to the early 2000s) has been attributed in part to the widespread abuse of steroids, although other factors are important as well. Steroids are a class of hormones derived from cholesterol and play a key role in sexual differentiation and reproduction. Examples include testosterone and estrogen, which play important roles in sexual differentiation in men and women, respectively (Hartgens and Kuipers 2004). Early research on anabolic steroid use in humans did not clearly reveal their full effects in the context of human performance (Cellotti and Cesi 1992), but more comprehensive reviews paint a more complete picture of their effects on human athletic performance (Bahrke and Yesalis 2004). Testosterone is especially interesting because of its role in building human muscle. During development, testosterone plays a key role in building male sexual structures and also assists in building bone, muscle, and cartilage. In adult male humans, treatment with exogenous testosterone has numerous effects; but, from the perspective of performance, perhaps the most important is skeletal muscle enlargement, which results in enhanced strength and potentially endurance, though this second effect is less well understood or proven. Less often discussed are the myriad number of side effects from taking exogenous testosterone, such as a weakened immune system, shrunken testicles, and an increased risk of cancer.

Testosterone is a naturally occurring hormone that plays a key role for male animals (including humans). In nonhuman animals, it exerts strong behavioral effects and levels of testosterone are often elevated when male animals compete for territories, for example (Marler and Moore 1988, 1989). It also appears to play a role in diminishing sensitivity to pain and increases aggression and risk taking. In humans, it has many of the same effects, although obviously in a different context (most of us rarely openly fight our neighbors for access to a house!). Thus, testosterone plays an important role that enables male animals to develop sexually and to exhibit typically male-like sexual behaviors, even if such behaviors can sometimes be deleterious. However, if levels of testosterone become elevated above normal levels for extended periods (weeks to months) in either humans or animals, these more natural effects can be exaggerated, and new effects can emerge.

In experiments with human volunteers, administration of exogenous testosterone dramatically increased overall strength at various tasks (e.g., amount of weight lifted during the leg press), primarily by increasing the size of individual muscle fibers, thus mimicking the effects of exercise (fig. 11.4). However, there are other effects on muscle tissue beyond an increase in fiber size. Testosterone also increases the total number of satellite cells, which can play a role in repairing the cell after damage. However, there is also an interactive effect with diet. In experiments with volunteers that were fed a high-protein diet and administered testosterone, there was an especially noticeable increase in muscle mass and strength. Interestingly,

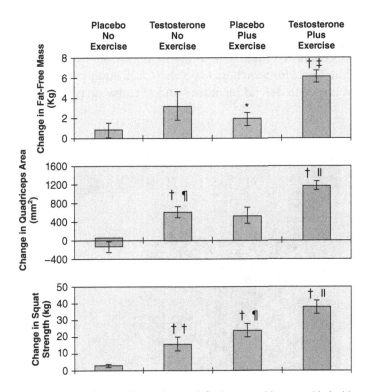

Fig. 11.4 Muscle mass and strength at a given task for human subjects provided with exogenous testosterone, as compared to that for control individuals. Note the dramatic increase in muscle mass and strength in testosterone-treated individuals, especially when combined with an exercise regimen. Taken from Bhasin et al. (2001) with permission from BioScientifica.

despite documented abuse of anabolic steroids for sports that emphasize endurance (e.g., bicycling), there is no proven benefit of anabolic steroids for endurance (Hartgens and Kuiper 2004), even though blood hemoglobin concentrations are affected.

However, it is also possible that this and other human laboratory studies are not fully testing the effects of anabolic steroids that athletes experience because of their short-term nature (Hartgens and Kuiper 2004), whereas some athletes are known to abuse steroids for long periods (years) of time. Further, athletes can employ dosing regimens 5–20 times higher than most laboratory studies (Hartgens and Kuiper 2004). Nonetheless, the above work seems to confirm the popular knowledge among steroid abusers that the best regimen for increasing muscle mass and strength is to combine the intake of steroids with the adoption of a high-protein diet.

In nonhuman animals, the addition of exogenous hormones, such as testosterone, seems to have a less obvious effect. In lizards, the addition of exogenous testosterone seems to improve maximum burst speed in at least a few lizard species, although in a survey of collared lizards, *Crotaphytus collaris*, there was no relationship between naturally occurring levels of testosterone levels and sprint speed (Husak et al. 2006). The addition of exogenous testosterone in *Gallotia* lizards enhanced the sizes of muscles in the head but had no apparent increase in bite force, for which this species is well equipped with its powerful jaws that it uses to defend territories and to consume hard prey (fig. 11.5;

Fig. 11.5 Image of a *Gallotia galloti* lizard. These lizards have hypertrophied jaws that they use in fights with other lizards to obtain resources. Exogenous treatment with testosterone caused further enlargement of their jaw muscles but did not result in improved bite performance. For more details, see the work by Huyghe et al. (2010). Image from Wikimedia Commons.

Huyghe et al. 2010). There is some evidence that elevated testosterone levels enhances endurance in lizards (Klukowski et al. 1998; Sinervo et al. 2000), a result that is intriguing given the lack of data for this effect in humans.

The fact that in at least one study (Huyghe et al. 2010) there was a mismatch between increased muscle size and performance is puzzling and suggests that there may be more complex underpinnings of performance in these and other animals. The impacts of training and diet in relation to animals and steroids remains an open topic, as issues of diet, stress, and different kinds of training all need to be examined further. It is also important to realize that seasonal shifts in levels of testosterone, as well as variation in testosterone levels among individual animals, play much more vital roles in many animals than they do in humans. A large body of work shows that, just as with humans, exogenous supplementation of testosterone exerts some powerful effects on behavior and mortality. In field studies with mountain spiny lizards, *Sceloporus jarrovi*, the addition of exogenous testosterone resulted in increased aggression and greater short-term territorial defense activity (e.g., displaying) but also increased mortality (Marler and Moore 1988, 1989). As with humans, highly elevated levels of testosterone can increase risk taking and aggression in some animals and, while such behaviors might offer some short-term benefits, they can also expose animals to predators or other risks. While modern medical companies will preach the value of "excess energy" to promote hormone replacement, in the natural world, high activity rates can also be dangerous. While much of the discussion has been on males, many of these same effects apply to female humans and animals, although this area has been less extensively studied. As an example, recent work with birds shows that exposure to elevated levels of testosterone (from a hormonal "challenge" via the injection of GnRH) resulted in female juncos, *Junco hyemalis*, spending less time brooding their nestlings, although they appeared to feed the nestlings more (Cain and Ketterson 2013). Within animals, therefore, there is a great deal we still don't know about the behavioral, physiological, and functional effects of elevated levels of hormones such as testosterone.

11.5 Are athletes anatomically different?

The notion of a close link between form and function in nonhuman animals has been demonstrated in many examples throughout this book. At least among well-defined species, variation in performance capacity is often closely tied to differences in anatomy. However, when one compares different individuals within a species, variation in morphology is often less tightly correlated with variation in performance. What about humans? As a single species, we exhibit remarkable variation in many traits, but is there any evidence that accomplished athletes are morphologically different from the rest of us? At first glance, the answer would seem to be yes, given that athletes are typically in elite condition; but, once the effects of training are removed, are there other, less noticeable aspects of their physiology that allow them to perform at high levels? This question is complex, but the modest amount of available data suggests that elite athletes in some sports seem to show genetic differences, morphological differences, or a combination, as compared to nonathletes.

The muscle architecture of elite sprinters seems to differ from that in elite marathon runners or in untrained athletes. Some evidence suggests that elite sprinters possess long muscle fascicles, and having such muscle this seems to be correlated with attaining higher maximum speeds during the 100 m dash, in a sample of male sprinters (Kumagai et al. 2000). Further, within female sprinters, there is also evidence for a correlation between muscle fascicle length and sprint speed. More recent research failed to find a link between lower leg musculoskeletal geometry and sprint performance, so this area requires further investigation (Karamanidis et al. 2011). In addition, athletic performance for activities such as sprinting is tied not only to muscle mass but also key joint angles that can enable individuals to have a favorable mechanical advantage, especially when the activity is over short time periods (e.g., the 100 m dash). For example, the Achilles tendon plays a key role in establishing the mechanical advantage of the foot during rapid sprinting. In a comparison of sprinters to nonsprinters, the Achilles tendon moment arms were about 25% smaller in sprinters compared to nonsprinters. Further, sprinters tended to have longer toes and shorter legs than nonsprinters did. In computer models, it was shown that this combination of traits enables sprinters to enjoy a superior "gear ratio" and thus to generate greater forward force (Lee and Piazza 2009). Therefore, there seems to be evidence that, for at least one activity, differences in muscle anatomy among individuals (athletes vs. nonathletes, and also comparing among athletes) translate into superior athletic performance. How such differences arise among individuals is a complex question for which there seem to be little data.

Recent work has also addressed the question of morphological correlates of elite performance in Kenyan marathon runners. In a comparison with Japanese runners, Kenyan runners exhibited an increased Achilles tendon moment arm and a decrease in the foot lever ratio (forefoot length divided by the Achilles tendon moment arm; Kunimassa et al. 2014). Although this morphology potentially provides additional stability, it reduces the storage and release of elastic energy and therefore decreases economy. This result conflicts with those from previous studies in which smaller muscle moment arms of the Achilles tendon were linked to elevated running economy and lower rates of metabolic energy consumption in highly trained runners (Scholz et al. 2008).

11.6 Extreme sports and human mortality

Unlike wild animals, humans have a penchant for seeking out extreme and often extremely dangerous sports, such as cave diving, parachuting off mountains, and high-elevation mountain climbing. While some of these sports do not involve extensive training or remarkable performance abilities (only a peaceful state of mind in a dangerous situation!), some extreme sports do require remarkable performance abilities. One example is extreme mountain climbing (climbing peaks >5,000 m), which involves unusually high levels of diffusion capacity within the lung. Such factors are acquired in part by intensive training, but there is also substantial genetic variation among humans in aerobic capacity. Nevertheless, even for humans in peak aerobic condition, mountain climbing is a singularly dangerous sport. While this fact has been known for years, much of the lore of mountain climbing has revolved around random factors being the principle causes of death and injury (i.e., random storms, a slip on ice above a ravine); however, recent research has dissected the factors that contribute to mortality on the highest peaks (Huey et al. 2007). The first and most obvious factor is mountain height; the process of attempting to climb high mountains imposes a higher mortality rate compared to that associated with climbing low mountains, particularly as climbers descend from the summit.

Indeed, it turns out that two of the most important factors that affect climber mortality are the age of the climber and whether the climber has actually summited (which is ironic, as the goal of mountain climbing is to reach the summit). Climbers that are older than 40 generally are less likely to reach the summit, and those older than 60 have significantly increased odds of dying (Huey et al. 2007; fig. 11.6). The age effect on climber mortality is intriguing, as some

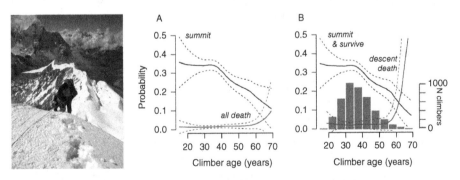

Fig. 11.6 *Left*, Image of a mountain climber. *Right*, Parameters associated with survival in mountain climbing. *A*, The link between climber age and the probability of summiting Mount Everest (1990–2005; black line) and of death (gray line) anywhere on the mountain. *B*, The likelihood of succeeding (black line: summiting and surviving) and of death (gray line) during descent. The histogram shows climber ages. Graph taken from Huey et al. (2007) with permission from the Royal Society Publishing Group. Image on left is from Wikimedia Commons.

evidence suggests that the kind of endurance that is used during strenuous mountain climbing declines less with age than the general pattern of aerobic performance does. However, the seemingly obvious effects of age are only part of the story; increased age was particularly deadly when in combination with descending from the summit. In other words, the worst combination for a mountain climber (from the perspective of surviving the climb) is both to be older and to actually reach the summit. This summit effect likely occurs because of extreme exhaustion (and perhaps in part to relaxation, which may cloud the ability to perceive and respond to dangers). In short, "selection" (in this case, a self-imposed form) is directional and clearly favors young mountaineers, particularly those that don't reach the summit. In other words, it's ok to turn around before reaching the summit!

Despite this result, there is some evidence that multiple lineages of humans originally evolved at somewhat higher elevations (1,000–2,000 m) before descending to lower elevations, and that, therefore, much of early human locomotor physiology evolved in a mild state of hypoxia. Further, the ancestral locomotor state of early humans is more akin to an intermittent, knuckle-walking, and less upright style than to the more adept and less intermittent walking and running styles of modern humans. This early evolution at relatively oxygen-poor higher altitudes also likely had significant energetic costs by limiting high levels of activity and therefore influencing both diet and brain size, as large brains require relatively high energy inputs.

One of the most prevalent aspects of human sports training consists of using "enhanced" methods that supposedly deliver improved performance results. For athletes that rely on aerobic endurance to win their chosen sports (e.g., bike racing, long-distance running), one aim is to improve VO_{2max}, a measure that defines the ability of the body to effectively deliver oxygen to fatigued muscles. Individuals with high values of VO_{2max} have a performance advantage during such sports because they can sustain aerobic metabolism (for hours) at a high speeds, as opposed to having to use anaerobic performance (which quickly exhausts muscles). Athletes have attempted a range of manipulations to increase VO_{2max}, including blood doping (removing blood, training, and then adding that blood back just prior to the race) and high-altitude training, among others. However, none of these methods significantly alters VO_{2max}; instead, they provide a short-term boost to endurance by providing more oxygen to the muscles, an effect which lasts only a few weeks. In short, the intrinsic locomotor physiology is not easily manipulated. This fact is sobering in light of the nearly 5,000-fold variation in VO_{2max} observed among animal species and suggests that the macroevolutionary patterns of locomotor physiology may be far more constrained that humans might care to admit.

References

Bahrke MS, Yesalis CE. 2004. Abuse of anabolic androgenic steroids and related substances in sport and exercise. Current Opinion in Pharmacology 4:614–620.

Bhasin S, Woodhouse L, Storer TW. 2001. Proof of the effect of testosterone on skeletal muscle. Journal of Endocrinology 170:27–38.

Bangsbo J, Norregaard L, Thorso F. 1991. Activity profile of competition soccer. Canadian Journal of Sport Sciences 16:110–116.

Cain KE, Ketterson ED. 2013. Individual variation in testosterone and parental care in a female songbird: The dark-eyed junco (*Junco hyemalis*). Hormones and Behavior 64:685–692.

Campos GE, Luecke TJ, Wendeln HK, Toma K, Hagerman FC, Murray TF, Ragg KE, Ratamess NA, Kraemer WJ, Staron RS. 2002. Muscular adaptations in response to three different resistance-training regimens: Specificity of repetition maximum training zones. European Journal of Applied Physiology 88:50–60.

Cellotti F, Cesi PN. 1992. Anabolic steroids: A review of their effects on the muscles, of their possible mechanisms of action and of their use in athletics. Journal of Steroid Biochemistry and Molecular Biology 43:469–477.

Garland T Jr, Else PL, Hulbert AJ, Tap P. 1987. Effects of endurance training and captivity on activity metabolism of lizards. American Journal of Physiology 252: R450–R456.

Gleeson TT. 1979. The effects of training and captivity on the metabolic capacity of the lizard *Sceloporus occidentalis*. Journal of Comparative Physiology B 129:123–128.

Goldspink DF. 1991. Exercise-related changes in protein turnover in mammalian striated muscle. Journal of Experimental Biology 160:127–148.

Hartgens F, Kuipers H. 2004. Effects of androgenic-anabolic steroids in athletes. Sports Medicine 34:513–554.

Hoppeler H, Fluck M. 2002. Normal mammalian skeletal muscle and its phenotypic plasticity. Journal of Experimental Biology 205:2143–2152.

Hoppeler H, Howald H, Conley K, Lindstedt SL, Claassen H, Vock P, Weibel ER. 1985. Endurance training in humans: Aerobic capacity and structure of skeletal muscle. Journal of Applied Physiology 59:320–327.

Huey RB, Salisbury R, Wang J-L, Mao M. 2007. Effects of age and gender on success and death of mountaineers on Mount Everest. Biology Letters 3:498–500.

Husak JF, Fox SF, Lovern MB, Van Den Bussche RA. 2006. Faster lizards sire more offspring: Sexual selection on whole-animal performance. Evolution 60:2122–2130.

Husak JF, Keith AR, Wittry BN. 2015. Making Olympic lizards: The effects of specialized exercise training on performance. Journal of Experimental Biology 218:899–906.

Huyghe K, Husak JF, Moore IT, Vanhooydonck B, Van Damme R, Molina-Borja M, Herrel A. 2010. Effects of testosterone on morphology, performance and muscle mass in a lizard. Journal of Experimental Zoology 313A:9–16.

Karamanidis K, Albracht K, Braunstein B, Catala MM, Goldmann JP, Bruggemann GP. 2011. Lower leg musculoskeletal geometry and sprint performance. Gait and Posture 34:138–141.

Klukowski M, Jenkinson NM, Nelson CE. 1998. Effects of testosterone on locomotor performance and growth in field-active northern fence lizards, *Sceloporus undulatus hyacinthinus*. Physiological Zoology 71:506–514.

Kumagai K, Abe T, Brechue WF, Ryushi T, Takano S, Mizuno M. 2000. Sprint performance is related to muscle fascicle length in male 100-m sprinters. Journal of Applied Physiology 88:811–816.

Kunimassa Y, Sano K, Oda T, Nicol C, Komi PV, Locatelli E, Ito A, Ishikawa M. 2014. Specific muscle-tendon architecture in elite Kenyan distance runners. Scandinavian Journal of Medicine and Science in Sports 24:e269–e274.

Lee SSM, Piazza SJ. 2009. Built for speed: Musculoskeletal structure and sprinting ability. Journal of Experimental Biology 212:3700–3707.

Marler CA, Moore MC. 1988. Evolutionary costs of aggression revealed by testosterone manipulations in free-living male lizards. Behavioral Ecology and Sociobiology 23:21–26.

Marler CA, Moore MC. 1989. Time and energy costs of aggression in testosterone-implanted free-living male mountain spiny lizards (*Sceloporus jarrovi*). Physiological Zoology 62:1334–1350.

O'Connor JL, McBrayer LD, Higham TE, Husak JF, Moore IT, Rostal DC. 2011. Effects of training and testosterone on muscle fiber types and locomotor performance in male six-lined racerunners (*Aspidoscelis sexlineata*). Physiological and Biochemical Zoology 84:394–405.

Scholz MN, Bobbert MF, van Soest AJ, Clark JR, Heerden J. 2008. Running biomechanics: Shorter heels, better economy. Journal of Experimental Biology 211:3266–3271.

Sinervo B, Miles DS, Frankino WA, Klukowski M, DeNardo DF. 2000. Testosterone, endurance, and Darwinian fitness: Natural and sexual selection on the physiological bases of alternative male behaviors in side-blotched lizards. Hormones and Behavior 38:222–233.

Van Damme R, Wilson R, Vanhooydonck B, Aerts P. 2002. Performance constraints in decathlon athletes. Nature 415:755–756.

Wang YX, Zhang CL, Yu RT, Cho HK, Nelson MC, Bayuga-Ocampo CR, Ham J, Kang H, Evans RM. 2004. Regulation of muscle fiber type and running endurance by PPARδ. PLoS Biology 2:e294.

Weinstein RB. 1995. Locomotor behavior of nocturnal ghost crabs on the beach: Focal animal sampling and instantaneous velocity from three-dimensional motion analysis. Journal of Experimental Biology 198:989–999.

Weinstein RB. 2001. Terrestrial intermittent exercise: Common issues for human athletics and comparative animal locomotion. Integrative and Comparative Biology 41:219–228.

Weinstein RB, Full RJ. 1999. Intermittent locomotion increases endurance in a gecko. Physiological and Biochemical Zoology 72:732–739.

Wilson M. 2013. Molecular signals and skeletal muscle adaptation to exercise. International Journal of Applied Exercise Physiology 2:1–10.

12 | Conclusion

We have examined how ecological and evolutionary forces have molded animal performance traits and, in doing so, we learned a great deal. Hopefully, the reader will take away several key messages from this synthesis. The first is that one cannot and should not attempt to decouple animal performance from the ecological and evolutionary context in which it resides. Every animal (including humans) exists within a broader ecological community that defines the limits of existence both in the presentation of threats (predation, starvation) and in the presentation of new opportunities (for reproduction, food, etc.). One of the emergent themes over the past several decades is that performance abilities make a profound difference in the lives of animals. At the most basic level, high performance capacities often mean that a frog, snake, mouse, or even a human (on a bad day!) may live for yet another day. Moreover, over the last several years, data have emerged suggesting a strong link between performance traits and sexual selection; high performance capacities may also mean high male quality, and therefore males with high performance capacities may be more likely to attain high reproductive success. So, in two of the most basic ways, superior performance at ecologically relevant tasks means that animals are more likely to both survive and breed, and this simple fact means that any comprehensive understanding of the demographic processes of animals should incorporate metrics of animal performance.

We have been impressed with how emerging portable technologies now enable humans to understand animal performance both in more natural circumstances (i.e., via the use of portable satellite tags and accelerometers) and in more precise and less intrusive ways than were possible before. We believe that this seamless integration of technology into the lives of animals is one of the most exciting and promising avenues for studying animal performance and animal movement. We also note that the increasing advent of both human

Animal Athletes. Duncan J. Irschick and Timothy E. Higham. © Duncan J. Irschick and Timothy E. Higham 2016. Published in 2016 by Oxford University Press.

culture and global warming onto ecological habitats and animal populations means that one of the most profound effects will be on patterns of animal movement and physiology, as animals are pushed into more remote and un-familiar habitats.

Another theme is the rise of evolutionary approaches for understanding both animal performance and the mechanisms that underlie it. The advent of phylogenetic techniques in the 1990s has now infiltrated itself into functional morphology, and evolutionary approaches are now fully appreciated as being complementary to mechanistic studies of animal function within single spe-cies. This evolutionary approach has been particularly useful for refuting the simplistic view that there is a 1:1 matching of morphology and function, and in turn, resource use. Rather, species within evolutionary clades and within ecological communities appear to have employed a variety of avenues to ac-complish the same set of functional tasks. Moreover, evolutionary studies have shown that single phenotypes are far more capable of accomplishing multiple functional tasks than previously believed.

Nevertheless, a strong conclusion from our book is that one should not ig-nore the important contribution of mechanistic studies for understanding both the limits on performance and the mechanisms that generating those limits. Research on ballistic movements, elastic components of muscles and tendons, and many other traits have shed new light onto enduring mysteries as to how animals can jump farther or move farther than previously thought possible. One of the most important principles emerging from studies of animal per-formance is that rules are meant to be broken. In some cases, the extreme per-formance capacities of animals are pointing the way for humans to develop new bioinspired technologies that may one day enable humans to achieve equally high levels of performance. However, before researchers proceed at breakneck speed to blindly distill the phenotypic traits of animals in the hopes of providing an instant performance boost, it is also important to realize that animals represent integrated structures with complex behaviors. The most successful attempts at bioinspiration take only the most salient elements of animal structure and design, and nothing else. The process of evolution, mani-fested over millions of years, has equipped animals with a range of structures and behavioral specializations for coping with extreme environmental condi-tions, and we are only beginning to understand the complex nature by which these components interact to influence performance. Thus, an integrative ap-proach may be our best hope for both understanding and appreciating nature, as well as employing it for our own purposes.

A final point relates to why one should care about performance traits at all; much of our natural attraction to animal performance is the same drive that compels us to compete in the human Olympics. Watching animals run after prey or crunch on bones is fun (at least for us!); this fascination can lead to an amusing game of lists, with a compilation of the "fastest" or "hardest-biting" animals, as well as to comparison to our more meager accomplishments. However, one of our messages is that such extreme performance traits evolved for a simple ecological need, and nothing more. In other words, the real world less resembles the human Olympics, in which everyone is fast and strong, and more resembles a highly constrained biological system in which the expression of every trait is only enough to survive. Constraints on performance, therefore, seem to be the rule more than the exception, and cases in which animals appear to have broken through such compromises in spectacular fashion are simply a reminder of how difficult it is to break basic rules of mechanics. The wonderful world of animal athletics therefore represents a world of opportunity for those of you interested in understanding the limits of evolution, and how animals can break through them. We hope that you will investigate them!

Artwork credit lines

Chapter	Figure/Table	Wiki permission
1	Figure 1.1A	By Fredlyfish4 (Own work) [CC BY-SA 3.0 (http://creativecommons.org/licenses/by-sa/3.0)], via Wikimedia Commons
1	Figure 1.3	Bundesarchiv, Bild 183-H0801-0020–001/CC-BY-SA [CC BY-SA 3.0 de (http://creativecommons.org/licenses/by-sa/3.0/de/deed.en)], via Wikimedia Commons
1	Figure 1.5	Pontia occidentalis 15780 by Walter Siegmund (talk)—Own work. Licensed under CC BY 2.5 via Wikimedia Commons—https://commons.wikimedia.org/wiki/File:Pontia_occidentalis_15780.JPG#/media/File:Pontia_occidentalis_15780.JPG
2	Figure 2.2	Selectiontypes-n0 images by Azcolvin429 (talk)—Selection Types Chart.png. Licensed under CC BY-SA 3.0 via Wikimedia Commons—http://commons.wikimedia.org/wiki/File:Selectiontypes-n0_images.png#/media/File:Selectiontypes-n0_images.png
2	Figure 2.8	Thamnophis ordinoides 2 by Will Brown—http://www.flickr.com/photos/37428634@N04/6965371158/. Licensed under CC BY 2.0 via Wikimedia Commons—https://commons.wikimedia.org/wiki/File:Thamnophis_ordinoides_2.jpg#/media/File:Thamnophis_ordinoides_2.jpg
3	Figure 3.1	By DesertStarSystems (DesertStarSystems) [CC BY 3.0 (http://creativecommons.org/licenses/by/3.0)], via Wikimedia Commons
3	Figure 3.1	By Phillip Colla (http://www.marinecsi.org/pop-up-satellite-tags/) [CC0], via Wikimedia Commons
3	Figure 3.1	By Tony Tucker [Attribution], via Wikimedia Commons
3	Figure 3.1	By Kristin Laidre [Public domain], via Wikimedia Commons
3	Figure 3.4	ZebraTailed-Lizard-640px. Licensed under Public Domain via Wikimedia Commons—https://commons.wikimedia.org/wiki/File:ZebraTailed-Lizard-640px.jpg#/media/File:ZebraTailed-Lizard-640px.jpg
3	Figure 3.5	By Mike Baird [CC BY 2.0 (http://creativecommons.org/licenses/by/2.0)], via Wikimedia Commons

Chapter	Figure/Table	Wiki permission
3	Figure 3.7	Bullshark Beqa Fiji 2007 by Pterantula. Licensed under CC BY 2.5 via Wikimedia Commons—http://commons.wikimedia.org/wiki/File:Bullshark_Beqa_Fiji_2007.jpg#mediaviewer/File:Bullshark_Beqa_Fiji_2007.jpg
4	Figure 4.3	Dolphins Swimming Near Catalina Island, CA by Jameyson MacDonald—Own work. Licensed under CC BY-SA 4.0 via Wikimedia Commons—https://commons.wikimedia.org/wiki/File:Dolphins_Swimming_Near_Catalina_Island,_CA.JPG#/media/File:Dolphins_Swimming_Near_Catalina_Island,_CA.JPG
4	Table 4.1	Greyhound Racing 4 amk by AngMoKio—Own work. Licensed under CC BY-SA 3.0 via Wikimedia Commons—https://commons.wikimedia.org/wiki/File:Greyhound_Racing_4_amk.jpg#/media/File:Greyhound_Racing_4_amk.jpg
5	Figure 5.1	By Wilhelm Ellenberger and Hermann Baum (University of Wisconsin Digital Collections[1]) [Public domain], via Wikimedia Commons
5	Figure 5.2	By NOAA CCMA Biogeography Team (http://www.photolib.noaa.gov/htmls/reef1841.htm) [Public domain], via Wikimedia Commons
5	Figure 5.7	Vanellus spinosus, Yeruham Park, israel by מינוזיג—Own work. Licensed under CC BY-SA 4.0 via Wikimedia Commons—http://commons.wikimedia.org/wiki/File:Vanellus_spinosus,_Yeruham_Park,_israel.jpg#mediaviewer/File:Vanellus_spinosus,_Yeruham_Park,_israel.jpg
5	Figure 5.7	Dragonfly hovering by Michael Palmer—Own work. Licensed under CC BY-SA 4.0 via Wikimedia Commons—http://commons.wikimedia.org/wiki/File:Dragonfly_hovering.jpg#mediaviewer/File:Dragonfly_hovering.jpg
5	Figure 5.7	PikiWiki Israel 11327 Wildlife and Plants of Israel-Bat-003 by Original photo: אורן פלס Oren PelesDerivative work: User:MathKnight—File:PikiWiki Israel 11327 Wildlife and Plants of Israel.JPG by פלס אורן. Licensed under CC BY 2.5 via Wikimedia Commons—http://commons.wikimedia.org/wiki/File:PikiWiki_Israel_11327_Wildlife_and_Plants_of_Israel-Bat-003.jpg#mediaviewer/File:PikiWiki_Israel_11327_Wildlife_and_Plants_of_Israel-Bat-003.jpg
5	Figure 5.8	Weisskopf Seeadler haliaeetus leucocephalus 2 amk by user:AngMoKio—selfmade photo at Deutsche Greifenwarte (Burg Guttenberg). Licensed under CC BY-SA 2.5 via Wikimedia Commons—https://commons.wikimedia.org/wiki/File:Weisskopf_Seeadler_haliaeetus_leucocephalus_2_amk.jpg#/media/File:Weisskopf_Seeadler_haliaeetus_leucocephalus_2_amk.jpg
5	Figure 5.8	Colibrí Cola de Oro (Golden-tailed Sapphire Hummingbird) Bigger File by Marcial4—Own work. Licensed under CC BY-SA 3.0 via Wikimedia Commons—https://commons.wikimedia.org/wiki/File:Colibr%C3%AD_Cola_de_Oro_(Golden-tailed_Sapphire_Hummingbird)_Bigger_File.jpg#/media/File:Colibr%C3%AD_Cola_de_Oro_(Golden-tailed_Sapphire_Hummingbird)_Bigger_File.jpg

Chapter	Figure/Table	Wiki permission
5	Figure 5.9	By Eva Rinaldi from Sydney Australia (Kangaroo hopping Uploaded by russavia) [CC BY-SA 2.0 (http://creativecommons.org/licenses/by-sa/2.0)], via Wikimedia Commons
6	Figure 6.3	PikiWiki Israel 11327 Wildlife and Plants of Israel-Bat-003 by Original photo: אורן פלס Oren PelesDerivative work: User:MathKnight—File:PikiWiki Israel 11327 Wildlife and Plants of Israel.JPG by פלס אורן. Licensed under CC BY 2.5 via Wikimedia Commons—https://commons.wikimedia.org/wiki/File:PikiWiki_Israel_11327_Wildlife_and_Plants_of_Israel-Bat-003.jpg#/media/File:PikiWiki_Israel_11327_Wildlife_and_Plants_of_Israel-Bat-003.jpg
6	Figure 6.6	Flying Dragon Mivart. Licensed under Public Domain via Wikimedia Commons—https://commons.wikimedia.org/wiki/File:Flying_Dragon_Mivart.png#/media/File:Flying_Dragon_Mivart.png
6	Figure 6.6	By Frederick P. Nodder [Public domain], via Wikimedia Commons
7	Figure 7.2	By Matthieu Deuté (Own work) [CC BY-SA 3.0 (http://creativecommons.org/licenses/by-sa/3.0) or GFDL (http://www.gnu.org/copyleft/fdl.html)], via Wikimedia Commons
9	Figure 9.3	By Nazir Amin (Mantis Shrimp) [CC BY-SA 2.0 (http://creativecommons.org/licenses/by-sa/2.0)], via Wikimedia Commons
9	Figure 9.4	By Dejean, A.; Delabie, J. H. C.; Corbara, B.; Azémar, F. D.; Groc, S.; Orivel, J. R. M.; Leponce, M. [CC BY 3.0 (http://creativecommons.org/licenses/by/3.0)], via Wikimedia Commons
9	Figure 9.7	Cascabelle. Licensed under Copyrighted free use via Wikimedia Commons—https://commons.wikimedia.org/wiki/File:Cascabelle.JPG#/media/File:Cascabelle.JPG
10	Figure 10.7	Thalassoma lunare 1 by Leonard Low from Australia—Flickr. Licensed under CC BY 2.0 via Wikimedia Commons—http://commons.wikimedia.org/wiki/File:Thalassoma_lunare_1.jpg#/media/File:Thalassoma_lunare_1.jpg
10	Figure 10.1	Men decathlon PV French Athletics Championships 2013 t142927 by © Marie-Lan Nguyen / Wikimedia Commons. Licensed under CC BY 3.0 via Wikimedia Commons—http://commons.wikimedia.org/wiki/File:Men_decathlon_PV_French_Athletics_Championships_2013_t142927.jpg#/media/File:Men_decathlon_PV_French_Athletics_Championships_2013_t142927.jpg
10	Figure 10.1	MichaelSmithDecathlon400meters by Template:Michael Smith—Own work. Licensed under Public Domain via Wikimedia Commons—http://commons.wikimedia.org/wiki/File:MichaelSmithDecathlon400meters.jpg#/media/File:MichaelSmithDecathlon400meters.jpg
10	Figure 10.1	Rico Freimuth 2012 Hypo-meeting by Erik van Leeuwen, attribution: Erik van Leeuwen (bron: Wikipedia).—http://www.erki.nl/pics/main.php?g2_itemId=67152. Licensed under GFDL via Wikimedia Commons—http://commons.wikimedia.org/wiki/File:Rico_Freimuth_2012_Hypo-meeting.jpg#/media/File:Rico_Freimuth_2012_Hypo-meeting.jpg

Chapter	Figure/Table	Wiki permission
11	Figure 11.5	Gallotia galloti cabeza by El fosilmaníaco—Own work. Licensed under CC BY-SA 3.0 via Wikimedia Commons—https://commons.wikimedia.org/wiki/File:Gallotia_galloti_cabeza.JPG#/media/File:Gallotia_galloti_cabeza.JPG
11	Figure 11.6	Summitting Island Peak by Mountaineer—Own work. Licensed under CC BY 3.0 via Wikimedia Commons—http://commons.wikimedia.org/wiki/File:Summitting_Island_Peak.jpg#/media/File:Summitting_Island_Peak.jpg

Index

Notes: Page numbers suffixed with *f* indicate figures, *t* indicate tables.

A

acceleration, rapid 5–6
acclimation,
        temperature 82–5
ACE 223
Achilles tendon 230
adaptive radiation 138
    definition 135–6
    performance
        evolution 135–9
adaptive traits 120
adhesive capacities 134–5
    incline locomotion *vs.* 156,
        157*f*, 158
    lizards 156–8
    running speed *vs.* 156
ADL (aerobic dive limit) 62
administration, exogenous
        see exogenous
        administration
aerobic capacity
    human athletics 221
    see also VO$_{2max}$
aerobic dive limit (ADL) 62
aerobic metabolism
    anaerobic *vs.* 71–3
    carbon dioxide
        production 71
    diving mammals 101–3
    endurance in human
        athletes 232
    VO$_{2max}$ 72–3, 72*f*
aerobic training 222
African finch (*Pyrenestes
        ostrinus*) 37–8*f*, 43
aggression, *testosterone* 226–7

amphibia
    frogs *see* frogs
    geographic variation in
        performance 208
    salamanders, tongue
        projection 20, 21*f*,
        106–7, 185–6
    toads *see* toads
anadromous cisco fish
        (*Coregonus artedii*) 207
anaerobic metabolism
    aerobic *vs.* 71–3
    efficiency 72
    extracellular potassium 71
    lactic acid buildup 71–2, 77
anatomical variations 96–8
    human athletics 229–30
    passing physical
        limit 103–8
Anderson, Chris 82
anoles (*Anolis*)
    adhesive capacities 134–5
    community ecology 212
    green anole lizards *see*
        green anole (*Anolis
        caroliensis*)
    jumping *see* anoles (*Anolis*),
        jumping
    movement variation 215,
        216*f*
    perching variation 212,
        213*f*
    sexual signals 168, 169*f*,
        170*f*
    tail loss locomotion
        effects 153–4, 153*f*

    temperature effects 87–8
    training effects 223–4, 224*f*
anoles (*Anolis*), jumping
        18–21, 18*f*
    behavioral adaptations 18
    habitat effects 20–1
    hindlimb length *vs.* takeoff
        velocity 20*f*
    morphological
        adaptations 18
    species effects 19, 20*f*
    stages 18–19, 19*f*
    types of jump 19–20, 20*f*
Anolis caroliensis see green
        anole (*Anolis
        caroliensis*)
Antilocapra americana
        (pronghorn) 75
ants
    adhesive capacities 135
    gliding 127, 128–9, 130*f*
    trap-jaw ants 187–9, 188*f*,
        189*f*
    waxrunner ants 135
arginine kinase
        (damselflies) 139
    burst speed 16–17
armaments, ornaments
        vs. 167–71
Armstrong, Lance 184
Arnold, Steve
    path analysis 26–7, 27*f*
    performance evolution 120
aspect ratio, butterfly
        wings 14
asymmetry, fluctuating 177

athletes *see* human athletes
Australian skinks, locomotion
  and temperature 87
Australian striped frogs
  (*Limnodynastes
  peroni*) 208
autonomy 152–3

B

Barry, Rick 17
basilisk *(Basiliscus)* 196, 197*f*
bats
  bite force quantification
    210, 210*f*
  bite performance 215, 217
  ecological niches 126
  morphological
    diversity 209–10, 209*f*
bat flight 124–6, 125*f*
  bird flight *vs.* 125–6
  diet 121
  evolution of 121
  kinematic flexibility 125
  plasticity 125
  wing movements 126*f*
  wing structure 122
beak size/shape 147–9
  biomechanics 211
  birdsong 149–50
  complex selection 37–8*f*, 43
  Darwin's finches 149
bearded dragon *(Pogna
  vitticeps)* 94
behavior
  Anolis jumping 18
  compensation *see*
    behavioral
    compensation
  ecological
    performance 50–2
  extreme
    performance 195–7
  performance
    relationship 17–21
  skewed distribution 12–13,
    12*f*
  thermoregulation 86–7
  trade-offs 160
behavioral
    compensation 62–4
  juveniles 63
  predator escape 63–4, 63*f*
biological system
    hierarchy 10–11

biomechanics 17
  Anolis jumping 18–19
  beak morphology 211
  closing structure 171
  insect flight 14
  performance trait 50
  swimming 152
  tongue projection 185, 186*f*
birds
  beak structure/size *see*
    beak size/shape
  crossbill finches *(Loxia
    curvirostra)* 37–8*f*, 43
  energy expenditure 60–1
  exogenous administration
    of testosterone 229
  flight *see* bird flight
  gyrfalcons *(Falco
    rusticolus)* 60–1
  hummingbirds *see*
    hummingbirds
  peregrine falcon *see*
    peregrine falcons
    (Falco perigrinus)
  sexual selection
    costs 171–2
  song *see* birdsong
bird flight
  bat flight *vs.* 125–6
  body pitch 123–4
  diving energetic costs 77
  evolution 121
  feathers 122
  maneuverability 121
  migration 123
  oxygen consumption in
    movement 69–70
  range of motion 122*f*
  sexual selection costs 171–
    2, 172*f*
  speed 121
birdsong 8, 149–50
bite force (strength) 93–4
  bats 210, 210*f*
  lizards 211
bluegill sunfish *(Lepomis
    macrochirus)* 154, 155*f*
body part loss 152–4
body pitch, bird flight 123–4
body shape, variability
    of 15–16
body size
  community ecology 211–12
  energetics of movement 69

flight 51, 51*f*
  power *vs.* in flight 107–8,
    108*f*
  rapid movements 185
body temperature control,
    ectotherms 85–7
Bolt, Usain 182
Bombina orientalis (Oriental
    toad) 40
bones
  in locomotion 96–7, 97*f*
  structural variation 98
brachiation 195–6
Brodie, Ed 43
Browning, Kurt 181
bull sharks *(Carcharhinus
    leucas)* 57–8, 58*f*
Bumpus, Herman 28
burrowing 109
  morphological
    adaptations 191–2
burst-speed locomotion
  damselfly escape
    strategies 16–17
  temperature effects in
    fishes 77*f*, 80
butterflies
  metabolic rate 204–5
  palatability in predation
    52
  population of origin
    studies 58, 59*f*
  speckled wood butterflies
    (*Pararge aegeria*) 203
  thermoregulation by
    wings 40
  wings 14–15, 15*f*
butterfly flight
  evolution of 14–15
  wing manipulation
    studies 39–40

C

caecilians 192
Calaveras Frog Jumping
    Contest 15
Callisaurus draconides (zebra-
    tailed lizard) 53–4, 55*f*
Callorhinus ursinus (northern
    seals) 102
carabid beetle *(Damaster
    blaptoides)* 151
Carassius (goldfish) 82–3
Carassius auratus 77*f*

carbon dioxide production, aerobic metabolism 71
Carchardon carcharias see great white shark (*Carchardon carcharias*)
Carcharhinus leucas (bull sharks) 57–8, 58*f*
Carcins maenas (shore crabs) 177
Caretta caretta (loggerhead turtles) 52
catapult mechanisms 111–12
cats, training 223
centrarchid fishes, suction feeding 133
chameleons (Chamaeleonidae)
hypoglossus muscle, temperature effects 81*f*, 82
tongue projection 110, 185
cheetahs (*Acinonyx jubatus*)
hunting studies 56–7, 57*f*
morphology *vs.* performance 35
Chrysopelea paradisi (flying snake) 129–30
chub mackerel (*Scomber japonicus*) 83*f*, 84
citrate synthase, heritability studies 202–3
Clark, Christopher 195
claws, biomechanics 171
clinging ability, geckos 81, 134, 156
cobweb spider (*Tidarren sisyphoides*) 159–60, 159*f*
cod (*Gadus morhua*) 95–6
collared lizard (*Crotaphytus collaris*) studies
exogenous administration of steroids 228–9
male competition 166
mark-recapture studies 33
performance effects 30, 32
survival/reproductive success studies 33, 34*f*
common iguana (*Iguana iguana*) 158
common-garden method, geographic variation in performance capacity 205–7

community ecology 200–19, 208–14
body size 211–12
coral reef fishes 212–13, 214*f*
direct interactions 201
limited resources 201
many-to-one and one-to-many mapping 214–17
morphological species difference 208–9
prey choice 210–11
comparative methods, performance evolution 118
constraints 143–62
ecological 158–60
physical laws 144
reproductive 158–60
see also innovations; trade-offs
convergent evolution 118, 120–30
morphology 121
coral reef fishes
community ecology 212–13, 214*f*
jaw size 215, 216*f*
core *temperatures*, shark species 100–1, 100*f*
Coregonus artedii (anadromous cisco fish) 207
crabs
claw trade-offs 151–2
fiddler crabs (*Uca pugnax*) 151–2
fighting stages 11
fluctuating asymmetry 177
ghost crabs 225
intermittent locomotion 225
shore crabs (*Carcins maenas*) 177
crayfish, dishonest sexual signaling 171, 171*f*
Crenicichla alta (pike cichlid) 41
crested *Anolis* lizards *see* anoles (*Anolis*)
cricket vocalization 205
critical thermal maximum temperature (CT_{max}) 85

crosbill finches (*Loxia curvirostra*) 37–8*f*, 43
Crotaphytus collaris see collared lizard (*Crotaphytus collaris*) studies
C-starts 5
Cyprinodon pecosensis (pupfish) 167

D

Damaster blaptoides (carabid beetle) 151
damselflies (*Enallagma*)
arginine kinase *see* arginine kinase (damselflies)
escape strategies 16–17
predator evasion 138–9
Darwin, Charles 2–3, 24, 117
Darwin's finches
beak size/shape 149
community ecology 211–12
decathletes 221–2, 222*f*
deer fighting, sexual selection 164–5
desert gecko (*Teratoscinus przewalski*) 76
desert lizard (*Sceloporus merriami*) 86–7
design, structure *vs.* 9
dewlap, lizards 168, 169*f*
diets 93
bat flight 121
Diplosaurus dorsalis 76*f*
directional selection 30, 30*f*, 31*f*
selection coefficient (beta) 35
displays, male competition 166
disruptive selection 30, 30*f*
distance running 221
distribution, skewed 12–13, 12*f*
diving flight
energetic costs 77
peregrine falcons (*Falco perigrinus*), diving speeds 190
speed, diving flight 190
speeds 51, 190
diving mammals
aerobic metabolism 101–3, 103*f*

diving mammals (*continued*)
 elephant seals *see* elephant
  seal diving
 extreme
  performance 182–3
 mitochondria 103*f*
 see also water diving
dolphins
 diving studies 48
 intermittent locomotion
  76, 76*f*
Draco (gliding lizard*)* 128,
 129*f*
Drosophia flight abilities
 extreme
  performance 196–7
 manipulative selection
  studies 42
Dudley, Robert 52
dung beetles
 male armaments 168–9
 male competition 167
 sexual selection costs 172
duress of animals, laboratory
  studies 47–8

E
ecological niches, bats 126
ecological performance 13–15
 behavior effects 50–2
 behavioral compensation
  see behavioral
   compensation
 definition 49–50
 emergent behaviors 61–2
 environmental effects 53–9
 maximum performance
  measure 49
 population of origin
  studies 58, 59*f*
ecology
 community *see* community
  ecology
 constraints 158–60
 performance *see* ecological
  performance
 physiological *see*
  physiological ecology
 specific traits *vs.* in natural
  selection 26
 temperature 85–8
ectotherms
 behavioral
  thermoregulation 86–7

body temperature
 control 85–7
 thermoconformers 86
El Niño 211–12
elephant seal diving 48
 extreme
  performance 182–3
 Northern elephant
  seals 56*f*
Emerson, Sharon 160
Enallagma see damselflies
 (*Enallagma*)
enclosed environment
 studies 40
Endler, John 24
endurance
 side-blotched lizards 32, 32*f*
 speed *vs.* in lizards 147, 148*f*
endurance training 222
energetics 68–70
 constraints on
  performance 21
 measurement 68–9
 variation in 73–6
energy expenditure, frog
  calls 4, 5*f*
energy requirements,
 morphology 144
environments, extreme *see*
  extreme environments
enzymes, temperature
  effect 76*f*, 77–8
EPOC (excess postexercise
  oxygen
  consumption) 77
estrogen 226
Eublepharis macularius
  (leopard gecko) 154
Eumetopias jubatus (Stellar sea
  lion) 102
Evans, Janet 17
evolution
 convergent *see* convergent
  evolution
 selective pressures 16
excess postexercise oxygen
  consumption
  (EPOC) 77
exogenous administration
 steroids 228–9
 testosterone 227–8, 227*f*
extracellular potassium,
  anaerobic
  metabolism 71

extreme environments 13–14
 performance evolution 139
extreme mountain
  climbing 231–2, 231*f*
extreme performance 181–99
 behavioral
  enhancement 195–7
 morphological
  mechanisms 192–5
 morphological
  novelties 190–2
 overcoming limits 182–4
 physiological
  mechanisms 192–5
 rapid movements *see* rapid
  movements
extreme sports 231–2

F
Falco perigrinus see peregrine
  falcons (*Falco
  perigrinus*)
Falco rusticolus
  (gyrfalcons) 60–1
fast-twitch muscle fibres 91,
  145, 145*f*
 slow-twitch *vs.* 192–3
feathers 122
feeding *see* diets
feeding structure, mechanical
  constraints 151–2
Felsenstein, Joseph 119
female choice, sexual
  selection 164, 174–7
fence lizard (*Sceloporus*)
  study 2
fiddler crabs (*Uca
  pugnax*) 151–2
field-portable
  instruments 48*f*
fighting, male
  competition 165–6,
  166–7
fish
 anadromous cisco fish
  (*Coregonus artedii*) 207
 bluegill sunfish (*Lepomis
  macrochirus*) 154, 155*f*
 bull sharks (*Carcharhinus
  leucas*) 57–8, 58*f*
 centrarchid fishes 133
 chub mackerel (*Scomber
  japonicus*) 83*f*, 84
 cod (*Gadus morhua*) 95–6

coral *reef see* coral reef
    fishes
female choice in sexual
    selection 176, 176*f*
fin manipulation
    studies 39
freezing survival 16
geographic variation in
    performance 207–8
goldfish (*Carassius*) 82–3
great white shark *see*
    great white shark
    (*Carchardon carcharias*)
guppies *see* guppies
    (*Poecilia reticulata*)
jaw structure 98, 98*f*, 136,
    136*f*, 137*f*, 138
killfish (*Rivulus hartii*) 146
labrid fishes (labridae) 136,
    136*f*, 137*f*, 138
largemouth bass
    (*Micropterus
    salmoides*) 154, 155*f*
mako shark *see* mako shark
    (*Isurus oxyrhincus*)
male competition 167
migration 146
mouth size variation 133
movement at different
    temperatures 83*f*, 84
muscle types 95–6
pharyngeal jaw 138
predator–prey
    relations 5–6
pupfish (*Cyprinodon
    pecosensis*) 167
red muscle 146, 147*f*
seahorses 132–3, 133*f*
silversides 207–8
speed of 52, 52*f*
stingrays 98, 98*f*
suction feeding 131, 131*f*,
    132–3, 132–4, 133, 133*f*,
    154, 155*f*
swimming structure–
    performance
    relation 152
tarpon (*Megalops
    atlanticus*) 57–8, 58*f*
temperature
    acclimation 82–3
temperature effects
    on burst speed
    locomotion 77*f*, 80

tracking studies in
    predator–prey
    relationships 57–8, 58*f*
wave action
    resistance 55–6
whale sharks (*Rhincodon
    typus*) 52, 52*f*
fitness
    definitions 27–8
    studies 24–46
    traits *vs.* measure 28
flapping flight, speeds 51
flea, locomotion 112
flight
    anatomical variation
        107–8, 107*f*
    body size effects 51, 51*f*
    diving *see* diving flight
    energy expenditure 50–1
    flapping flight speeds 51
    fluid dynamics 186–7
    independent
        evolution 121–7
        see also bat flight; bird
        flight; insect flight
    manipulation
        studies 39–40
    maximum speeds 61
    minimum energy
        expenditure 51
    power *vs.* speed 107–8
    styles of 51
    tail wind effects 53
flight muscles, temperature
    control in insects 86
fluctuating asymmetry 177
fluid dynamics 186–7
flying frogs 11, 160
flying snake (*Chrysopelea
    paradisi*) 129–30
Fosbury, Dick 9, 184
Fosbury flop 9–10, 9*f*, 184
freezing survival, fish 16
frequency *bandwidth*,
    birdsong 149–50, 150*f*
frogs
    calling 4, 8
    calling in sexual selection
        costs 173, 174
    elastic mechanisms in
        locomotion 111–12
    extreme performance in
        jumping 183
    feeding modes 104*f*

flying frogs 11, 160
geographic variation in
    performance 208
gliding 127
jumping competitions 50–2
Q_{10} and temperature 82
suction feeding 132
tongue projection 20, 103–4
tree frogs (*Hyla crucifer*) 82
function, performance *vs.* 8–9
Fundulus heteroclitus 77*f*

G
Gadus morhua (cod) 95–6
Galapagos island birds *see*
    Darwin's finches
Gallotia lizards 228*f*
    exogenous administration
        of steroids 228–9
gape duration, lizards 94–5
garden snails (*Helis
    aspersa*) 32–3
garter snakes (*Thamnophis*)
    complex selection on
        locomotion 43, 44*f*
    heritability studies 202–3
Gasterosteus aculeatus (three-
    spine stickleback) 207
geckos
    clinging ability 81, 134, 156
    desert gecko (*Teratoscinus
        przewalski*) 76
    distance capacity 225–6,
        225*f*
    Gehyra vorax. 135*f*
    leopard gecko (*Eublepharis
        macularius*) 154
    Pachydactylus 156–8
    running speed 60
    Tokay gecko (*Gekko
        gecko*) 135*f*
Gehyra vorax. 135*f*
Gekko gecko (Tokay
    gecko) 135*f*
genetics 202–5
    heritability 202–3
    random genetic drift 25
    sexual selection 165
geographic variation
    in performance
        capacity 201*f*, 205–8
    common-garden
        method 205–7
    predator pressure 206–7

Gerrhonotus multicarinatus 76*f*
ghost crabs 225
Gleeson, Todd 2
gliding 127–30
  multiple evolution 127
gliding lizard *(Draco)* 128,
  129*f*
goldfish *(Carassius)* 82–3
Gould, Steven Jay 143
GPS tracking devices 56–7,
  57*f*
Grant, Bruce 86–7
Grant, Peter 211–12
Grant, Rosemary 211–12
great white shark *(Carchardon
  carcharias)* 99*f*
  performance in suboptimal
    settings 99–101
green anole *(Anolis caroliensis)*
  female choice in sexual
    selection 176–7
  male competition 167
ground squirrels 84
gular pump 73, 74*f*, 75, 75*f*
guppies *(Poecilia reticulata)*
  burst-speed performance
    206–7, 206*t*
  female choice in sexual
    selection 176, 176*f*
  predator–prey
    interactions 41
gyrfalcons *(Falco
  rusticolus)* 60–1

H
habitats
  Anolis jumping 20–1
  damselfly burst speed 17
handicap hypothesis, sexual
  signals 168
head-bobbing 166
Helis aspersa (garden
  snails) 32–3
Hendenström, Anders 125–6
heritability 11–13
  genetics 202–3
  insects 203–4, 204*f*
  traits 13
Higham, Tim 154
hindlimb length
  Anolis jumping 20*f*
  Urosaurus lizards 38
homoplasty 118
horned lizards 94

horses
  gait and energetics 70
  heritability 203
human athletes 220–32
  aerobic capacity 221
  aerobic endurance 232
  anatomical
    differences 229–30
  animals *vs.* 220–1
  distance running 221
  extreme sports 231–2
  specialists *vs.*
    generalists 221–2
  training 222–6
  VO$_{2max}$ 232
hummingbirds
  extreme environments 139
  extreme locomotion 195
Husak, Jerry 30, 32
  survival/reproductive
    success studies 33
hydrostatic elongation,
  tongue projection 104
Hygrolycosa rubrofasciata (wolf
  spider) 174, 175*f*
Hyla crucifer (tree frogs) 82
Hyla microcephala 5*f*

I
Iguana iguana (common
  iguana) 158
IMU *(internal measurement
  unit)*, cheetah hunting,
  57 57*f*
incline locomotion 53–4, 55*f*
  adhesion *vs.* 156, 157*f*, 158
  downhill 54
  trade-offs 156, 158
inertial elongation, tongue
  projection 104
innovations 155–8
  definition 155–6
  see also constraints;
    trade-offs
insect(s)
  adhesive qualities 134
  ants *see* ants
  butterflies *see* butterflies
  carabid beetle *(Damaster
    blaptoides)* 151
  cobweb spider *(Tidarren
    sisyphoides)* 159–60, 159*f*
  damselflies *see* damselflies
    (Enallagma)

Drosophila see Drosophia
  flight abilities
  feeding phenotypes 151
  flea locomotion 112
  flight *see* insect flight
  gliding 127, 128–9, 130*f*
  heritability 203–4
  leafhopper insect
    locomotion 111–12
  metabolic rate
    heritability 204–5
  sexual dimorphism 159–
    60, 159*f*
  sexual selection costs 172,
    174, 175*f*
  stalk-eyed fly 172, 174
  temperature control 86–7
  tree crickets 174
  wolf spider *(Hygrolycosa
    rubrofasciata)* 174, 175*f*
insect flight 126–7
  biomechanics 14
  ecological performance 49
  evolution of 121
  extreme
    performance 196–7
  size effects 121–2
  Troponin-t 203–4, 204*f*
  wing movements 126–7,
    127*f*
insectivores, mammals *vs.*
  reptiles 94
intermittent
  locomotion 224–6
  energy expenditure 75–6,
    76*f*
  manipulative selection
    studies 41–2
internal measurement unit
  (IMU), cheetah
  hunting 57, 57*f*
invertebrates
  jumping 189–90
  male armaments 168–9
  see also insect(s)
inverted-pendulum motion,
  hip movement 112–13
islands, community
  ecology 211–12
Isurus oxyrhincus see
  mako shark *(Isurus
  oxyrhincus)*
Italian wall lizards *(Podalcis
  sicula)* 41

J

jaws 93–4
  biomechanics 171
  coral reef fishes 215, 216f
  pharyngeal see pharyngeal
    jaws
  stingrays 98, 98f
  studies of 118
  trade-offs 151
jumping
  Anolis lizards see anoles
    (Anolis), jumping
  frogs 50–2, 183
  invertebrates 189–90
  kangaroos 111, 112–13, 112f
juveniles, behavioral
  compensation 63

K

kangaroo jumping 111,
  112–13, 112f
key joint angles, human
  sprinters 230
killfish (Rivulus hartii) 146
Kimetto, Dennis 221
kinematic flexibility, bat
  flight 125
Kingsolver, Joel 29–30, 31f
  selection coefficient (beta)
    tests 35–8
Komodo dragon see Varanid
  lizards

L

laboratory studies
  control in 47
  duress of animals 47–8
  single behaviors 49
  studies in nature 47–9
  temperature
    acclimation 84–5
labrid fish (labridae) 136,
  136f, 137f, 138
lacertid lizards
  fluctuating asymmetry
    178
  muscle types 95
lactate dehydrogenase,
  damselflies
    (Enallagma) 139
  burst speed 16–17
lactic acid, anaerobic
  metabolism 71–2, 77
Lailvaux, Simon 205

Lamna ditropis (salmon
  shark) 100f, 101
Lampris guttatus (opah) 92
largemouth bass (Micropterus
  salmoides) 154, 155f
leafhopper insect
  locomotion 111–12
leopard gecko (Eublepharis
  macularius) 154
Leptonychotes weddellii
  (Weddel seal) 102–3
Lewis, Carl 190
Lewontin, Richard 143
limited resources, community
  ecology 201
limits, conception of 183
Limnodynastes peroni
  (Australian striped
    frogs) 208
linkage disequilibria 25
lizards
  adhesive capacities 134–5,
    134f, 135f, 156–8
  adhesive feet see adhesive
    capacities
  Anolis see anoles (Anolis)
  Australian skinks 87
  bearded dragon (Pogna
    vitticeps) 94
  behavioral
    thermoregulation
    studies 86–7
  bite force
    quantification 211
  chameleon see chameleons
    (Chamaeleonidae)
  collard lizards see collared
    lizard (Crotaphytus
      collaris) studies
  common iguana (Iguana
    iguana) 158
  community ecology 212
  desert gecko (Teratoscinus
    przewalski) 76
  desert lizard (Sceloporus
    merriami) 86–7
  egg mass 158
  exogenous administration
    of steroids 228–9
  female choice in sexual
    selection 176–7
  fence lizard (Sceloporus)
    study 2
  fluctuating asymmetry 178

Gallotia see Gallotia lizards
gape duration 94–5
geckos (Pachydactylus)
  156–8
gliding lizard (Draco) 128,
  129f
hindlimb length studies 38
incline locomotion 53–4, 55f
innovations 156–8
intermittent locomotion 76,
  224–5, 225f
jaw strength 94
Komodo dragon see
  Varanid lizards
lacertid lizards see lacertid
  lizards
leopard gecko (Eublepharis
  macularius) 154
locomotion and
  temperature 87
male competition 166
maximum capacity in
  predator escape
  63–4, 63f
mountain spiny lizards
  (Sceloporus jarrovi) 229
predator escape 60
prey choice 210–11
reproductive
  constraints 158
sexual signals 168, 169f
speed vs. endurance 147,
  148f
tail loss 153–4, 153f
tail loss locomotion
  effects 154
temperature effects 76, 76f
Urosaurus lizards 38
Varanid lizards see Varanid
  lizards
wall-climbing see adhesive
  capacities
zebra-tailed lizard
  (Callisaurus
    draconides) 53–4, 55f
locomotion
  anatomical variations
    96–7, 97f
  behavior effects on
    performance 50
  burst-speed see burst-speed
    locomotion
  complex selection 43
  elastic mechanisms 110–12

locomotion (*continued*)
 endurance in male
   competition 166
 inclines *see* incline
   locomotion
 on inclines *see* incline
   locomotion
 intermittent *see* intermittent
   locomotion
 tetrapods 96
 toads 10
loggerhead turtles (*Caretta
   caretta*) 52
Loxia curvirostra (crosbill
   finches) 37–8*f*, 43

M
Macropus eugenii (tammar
   wallaby) 113
mako shark (*Isurus
   oxyrhincus*) 99*f*
 muscle anatomy 101
 performance in suboptimal
   settings 99–101
male armaments,
   invertebrates 168–9
male competition, sexual
   selection 163, 164,
   165–7
mammals
 bats *see* bats
 breathing efficiency 73
 cheetahs *see* cheetahs
   (*Acinonyx jubatus*)
 dolphins *see* dolphins
 gliding 127
 ground squirrels 84
 horses *see* horses
 kangaroo jumping 111,
   112–13, 112*f*
 locomotion 113
 marine *see* marine
   mammals
 pronghorn (*Antilocapra
   americana*) 75
 rodent training 223
 tammar wallaby (*Macropus
   eugenii*) 113
 temperature acclimation 84
 VO$_{2max}$ 73, 74*t*, 75
 wallabies 113
mandibular raking 109–10
maneuverability, bird
   flight 121

manipulative selection
   studies 38–41
 butterfly flight 39–40
 Drosophia flight abilities 42
 intermittent
   locomotion 41–2
 Oriental toad (*Bombina
   orientalis*) 40
 predator–prey
   interactions 41
mantis shrimps 187–8, 187*f*
many-to-one mapping,
   community
   ecology 214–17
Marden, Rick 203–4, 204*f*
marine mammals
 aerobic dive limit 62
 aerobic metabolism 102–3
 diving energetic costs 77
 elephant seals *see* elephant
   seal diving
 northern seals (*Callorhinus
   ursinus*) 102
 suction feeding 131
 water diving 54
 Weddel seal (*Leptonychotes
   weddellii*) 102–3
mark-recapture studies 28–9
 collared lizards (*Crotaphytus
   collaris*) 33
 morphology *vs.*
   performance 35
 performance effects on
   fitness 29
 performance limitation
   studies 33–4
 survival/reproductive
   success studies 33
mass-spring model, elastic
   locomotion 112
mating, reproductive
   constraints 158–9
maximum capacity 59–61
maximum performance
   measure, ecological
   performance 49
maximum power
   output 144–5
McPeek, Mark 16
mechanical constraints,
   feeding
   structure 151–2
mechanical pulling, tongue
   projection 104

mechanical trade-offs 144
mechanistic constraints on
   performance 21
medial tympaniform
   membrane 149
Megalops atlanticus
   (tarpon) 57–8, 58*f*
metabolic rate depression,
   turtles 77–8
metabolism
 aerobic *see* aerobic
   metabolism
 anaerobic *see* anaerobic
   metabolism
mice (*Mus domesticus*) 41–2
Micropterus salmoides
   (largemouth bass)
   154, 155*f*
migration
 bird flight 123
 fish 146
 geographic variation in
   performance 207
minimum energy
   expenditure, flight 51
Mirounga angustirostris
   (Northern elephant
   seals) 56*f*
mitochondria 103*f*
moray eel (*Muraena
   retifera*) 190–1, 191*f*
morphology
 Anolis jumping 18
 convergent evolution 121
 energy requirements 144
 extreme performance
   190–2, 192–5
 nonlinear effects 10–11
 performance *vs.* in
   selection 35–8
 see also selection
   coefficient (beta)
morphs, male
   competition 167
mortality, extreme mountain
   climbing 231–2, 231*f*
motor control, tongue
   projection 106
mountain climbing,
   extreme 231–2, 231*f*
mountain spiny lizards
   (*Sceloporus jarrovi*) 229
mouth size variation,
   fish 133

movement
  elastic mechanisms 110–12
  energetics 68–70
  harmful byproducts 76–8
  rapid *see* rapid movements
  variation in anole lizards
    (*Anolis caroliensis*) 215,
    216*f*
  see also locomotion
Muraena retifera (moray
  eel) 190–1, 191*f*
Mus domesticus (mice) 41–2
muscle
  cod (*Gadus morhua*) 95–6
  fast-twitch fibres *see* fast-
    twitch muscle fibres
  fibres 91
  human sprinters 230
  lacertid lizards 95
  in locomotion 96–7, 97*f*
  mechanism of action 21
  pelagic fish 95
  performance in suboptimal
    settings 99
  red *see* red muscle
  slow-twitch *see* slow-twitch
    muscle fibres
  temperature effects
    on 80–2
  testosterone 226
  tongue projection 106
  types of 97–8
mutations, spread 200
Myoxocephalus scorpulus 77*f*

N
natural selection 2–3, 24–5
  enclosed environment
    studies 40
  impact quantification
    28–9
  individual variation 24
  mechanism of action 4
  multiple causes of 25
  over-population 24
  path diagrams 25–7
  performance effects, lack of
    evidence 34–5
  specific traits *vs.* ecological
    context 26
noise production, rattlesnakes
  192, 193*f*, 194*f*
nonlinear effects,
  morphology 10–11

non-performance traits,
  performance traits
  vs. 8
normal distribution,
  structure 12–13, 12*f*
Northern elephant
  seals (*Mirounga
  angustirostris*) 56*f*
northern seals (*Callorhinus
  ursinus*) 102
novel capacities, performance
  evolution 130–5
novel habitats, performance
  evolution 138–9
Nyad, Diana 182

O
offspring care, female
  choice in sexual
  selection 175–6
one-to-many mapping,
  community
  ecology 214–17
on-water locomotion 196,
  197*f*
opah (*Lampris guttatus*) 92
orbatid mites 181
Oriental toad (*Bombina
  orientalis*) 40
ornaments, armaments
  vs. 167–71
over-population, natural
  selection 24
oxygen use, energetics of
  movement 68–9

P
Pachydactylus 156–8
pain sensitivity,
  testosterone 226–7
Pararge aegeria (speckled
  wood butterflies) 203
particle image velocimetry
  (PIV) 123, 124*f*
Patek, Sheila 187–8
paternity, fitness
  definition 27
path analysis 27*f*
  example 26–7
path diagrams
  multiple trait analysis 26
  natural selection 25–7
peacock, sexual selection 163
pelagic fish, muscle types 95

perching variation,
  Anolis 212, 213*f*
peregrine falcons (*Falco
  perigrinus*)
  diving speeds 190
  energy expenditure 60–1
  extreme performance
    182–3
performance
  capacity 10
  definitions 1–2, 6–9
  ecological *see* ecological
    performance
  ecological context *see*
    ecological performance
  effects *see* performance
    effects
  energetic/mechanistic
    constraints 21
  enhancers 183–4, 184*f*
  environmental effects 67–8
  evolution *see* performance
    evolution
  extreme *see* extreme
    performance
  function *vs.* 8–9
  geographic variation
    in capacity *see*
    geographic variation
    in performance
    capacity
  high specialization
    108–10
  limitation studies 33–4
  manipulative studies *see*
    manipulative selection
    studies
  measures in hierarchical
    structure 11
  mechanical constraints 21,
    147–55
  mechanistic limits 91–2
  morphology *vs.* in
    selection 35–8
  see also selection
    coefficient (beta)
  natural selection, lack of
    evidence 34–5
  in nature *see* performance
    in nature
  obsession with 3
  quantitative measures 8–9
  sexual selection
    costs 171–4

performance (*continued*)
  suboptimal settings *see*
    performance in
    suboptimal settings
  symmetry effects 177–8
  temperature effects 87–8
  traits *see* performance traits
  whole-organism view *see*
    whole-organism view
    of performance
performance effects 29–35
  collared lizard (*Crotaphytus
    collaris*) studies 30, 32
  mark-recapture studies 29
  stabilizing selection 29–30,
    30*f*
performance evolution
    15–17, 91–142
  adaptive radiation 135–9
  adaptive traits 120
  comparative methods 118
  extreme environments 139
  methods 117–19
  models 120
  novel capacities 130–5
  novel habitats 138–9
  quantitative analysis of
    data 119
  see also convergent
    evolution
performance in nature 47–66
  reasons for study 47–9
performance in suboptimal
    settings 99–103
  muscles 99
performance traits
  non-performance traits
    vs. 8
  repeatability 13
Pgi 204–5
pharyngeal jaws 132, 138
  moray eel 190–1, 191*f*
Phelps, Michael 182
phosphoglucose isomerase
    gene 204–5
physical laws,
    constraints 144
physiological ecology 67–90
  aerobic *vs.* anaerobic
    metabolism 71–3
  definition 68
  energetics *see* energetics
  movement *see* movement
    movement
    byproducts 76–8

movement energetics 73–6
  temperature
    acclimation 82–5
  temperature effects 78–82
physiological enhancers 184
physiological mechanisms,
    extreme
    performance 192–5
pike cichlid (*Crenicichla
    alta*) 41
PIV (particle image
    velocimetry) 123, 124*f*
plasticity, bat flight 125
pleiotropic responses 25
Podalcis sicula (Italian wall
    lizards) 41
Poecilia reticulata see guppies
    (*Poecilia reticulata*)
populations 200–1
  origin studies 58, 59*f*
potassium, extracellular 71
power
  body size *vs.* in flight 107–
    8, 108*f*
  speed *vs.* in flight 107–8
predator escape/evasion 50
  complex selection 43, 44*f*
  energy expenditure 60
  locomotion speed 52
  maximum capacity 63–4,
    63*f*
  non-locomotion
    methods 52
  performance
    evolution 138–9
predator pressure,
    performance
    capacity 206–7
predator–prey
    interactions 5–6
  ecological
    performance 56–7
  fish 5–6
  manipulative selection
    studies 41
  prey grabbing 103–4
  prey size 109–10
  tracking studies 57–8, 58*f*
prey choice, community
    ecology 210–11
pronghorn (*Antilocapra
    americana*) 75
pupfish (*Cyprinodon
    pecosensis*) 167
pushups 166

Pyrenestes ostrinus (African
    finch) 37–8*f*, 43
pyruvate kinase
    (damselflies) 139
  burst speed 16–17

Q
Q_{10}, temperature effects 82
quantification of tasks 6
quantitative measures,
    performance 8–9

R
radiation, adaptive *see*
    adaptive radiation
random genetic drift 25
range of motion, bird
    flight 122*f*
rapid acceleration, predator–
    prey relations 5–6
rapid movements 184–90
  body size 185
  tongue projection 185–6
rattlesnakes 192, 193*f*, 194–5,
    194*f*
red muscle 95
  distribution of 146
  fish 146, 147*f*
relative loading of a
    system 14
repeatability 11–13
  performance traits 13
reproduction
  constraints 158–60
  stabilizing selection 32–3
  success studies 33, 34*f*
reptiles
  ambushing strategies 73
  breathing efficiency 73
  lizards *see* lizards
  snakes *see* snakes
  squamate reptiles 223
  training 223–4
  turtles *see* turtles
  VO_{2max} 73
rete mirabile 100–1
Rezazadeh, Hossein 181
Reznick, David
  guppy predator
    pressure 206–7, 206*t*
  predator–prey
    interactions 41
Rhincodon typus (whale
    sharks) 52, 52*f*
Riskin, Dan

bat flight 125
bat *vs.* bird flight 125–6
Rivulus hartii (killfish) 146
Roberts, Thomas 183
rodent training 223
running, distance 221
running speed
    adhesive capacities *vs.* 156
    male competition 166

S
salamanders, tongue
    projection 20, 21*f*,
    106–7, 185–6
salmon shark *(Lamna
    ditropis)* 100*f*, 101
sample sizes, statistical
    methods 29
Sceloporus (fence lizard) 2
Sceloporus jarrovi (mountain
    spiny lizards) 229
Sceloporus merriami (desert
    lizard) 86–7
Sceloporus undulatus 76*f*
Scomber japonicus (chub
    mackerel) 83*f*, 84
sea horses, suction
    feeding 132–3, 133*f*
sea turtles, speed of 52
selection
    directional *see* directional
        selection
    disruptive 30, 30*f*
    manipulative studies *see*
        manipulative selection
        studies
    morphology *vs.*
        performance 35–8
        see also selection
        coefficient (beta)
    natural *see* natural selection
    sexual *see* sexual selection
    stabilizing *see* stabilizing
        selection
selection coefficient (beta) 35–8
    tests for 36*f*, 37*f*
selective breeding
    programs 203
selective pressures,
    evolution 16
sequential assessment game,
    male competition 166
sexual selection 4, 24–5,
    159–60, 163–80
    definition 163–4

female choice 164, 174–7
    functional approach 164–5
    genetic quality 165
    impact quantification 28–9
    male competition 163, 164,
    165–7
    performance costs 171–4
    see also sexual signals
sexual signals 163, 165, 168
    armaments *vs.*
        ornaments 167–71
    dishonest signaling 169–71
    handicap hypothesis 168
    honesty of 167–71
sexual size dimorphism
    159–60, 159*f*
shore crabs *(Carcins
    maenas)* 177
side-blotch lizard *(Uta
    stansburiana)*
    egg mass 158
    endurance capacity 32, 32*f*
    male competition 167
    reproduction in stabilizing
        selection 32, 32*f*
sidewinding 195–6
signaling, sexual *see* sexual
    signals
silverside fish 207–8
size reduction, butterfly
    wings 14–15, 15*f*
skewed distribution,
    behavior 12–13, 12*f*
skin flap enlargement 166
slow-twitch muscle fibres 91
    fast-twitch muscle fibres
        vs. 192–3
temperature
    acclimation 83–4
snakes
    flying snake *(Chrysopelea
        paradisi)* 129–30
    garter snakes *see* garter
        snakes *(Thamnophis)*
    gliding 127
    noise production 192, 193*f*,
        194–5, 194*f*
    rattlesnakes 192, 193*f*,
        194–5, 194*f*
soccer players, intermittent
    locomotion 225–6
soft tissues, in
    locomotion 96–7, 97*f*
space limitations, body
    structure 144

specific traits, ecological
    context *vs.* in natural
    selection 26
speckled wood butterflies
    (Pararge aegeria) 203
speed
    bird flight 121
    diving flight 190
    endurance *vs.* in
        lizards 147, 148*f*
    energetics of
        movement 69–70
    gait type 70
    power *vs.* in flight
        107–8
sphinx moths 86
sprint speed studies
    lizard tail loss 153–4
    Urosaurus lizards 38
squamate reptiles,
    training 223
stabilizing selection 30, 30*f*,
    31*f*, 32–3
    performance effects 29–30,
        30*f*
    selection coefficient
        (beta) 35
stalk-eyed fly 172, 174
statistical methods 26
    multiple trait analysis *see*
        path diagrams
    sample sizes 29
Stellar sea lion *(Eumetopias
    jubatus)* 102
steroids 226–9
    definition 226
    exogenous administration
        228–9
stingrays 98, 98*f*
stress hormones 224
structure
    design *vs.* 9
    normal distribution 12–13,
        12*f*
suction feeding 131–4
    fishes 131, 131*f*, 132–4,
        133*f*, 154, 155*f*
    frogs 132
    marine mammals 131
    quantification 132–3
    sea horses 132–3, 133*f*
survival
    fitness definition 27
    reproductive success and,
        studies of 33

survival success studies 33, 34*f*
Swartz, Sharon 125–6
swimming
  biomechanics 152
  energetics of movement 70
  fluid dynamics 186–7
symmetry effects, performance 177–8
syrinx, birdsong 149
system, relative loading of a 14

T
tail loss, lizards 153–4, 153*f*
takeoff velocity, *Anolis* jumping 20*f*
tammar wallaby (*Macropus eugenii*) 113
tarpon (*Megalops atlanticus*) 57–8, 58*f*
task quanitifcation 6
Taurulus bubalis 77*f*
Taylor, Charles 70
temperature 67
  acclimation *see* temperature acclimation
  burst speed locomotion 77*f*
  deleterious effects 85
    see also critical thermal maximum temperature (CT$_{max}$)
  ecological context 85–8
  on muscular properties 80–2
  physiological ecology 78–82
temperature acclimation 82–5
  physiological effects 84–5
  single temperature only 84
tendons, in locomotion 96–7, 97*f*, 111–12
Teratoscinus przewalski (desert gecko) 76
testosterone 226–8
  behavior effects 226–7
  exogenous administration 227–8, 227*f*, 229
  seasonal shifts 229
  survival/reproductive success 33

tetrapods, locomotion 96
Thamnophis see garter snakes (*Thamnophis*)
thermoconformers 86
thermoregulation, butterfly wings 14–15, 40
threadsnakes, feeding strategies 109–10
three-spine stickleback (*Gasterosteus aculeatus*) 207
Tidarren sisyphoides (cobweb spider) 159–60, 159*f*
toadfish, calls 192, 193–4, 194*f*
toads
  enclosed environment studies 40
  locomotor ability 10
  Oriental toad (*Bombina orientalis*) 40
  tongue projection 10, 104
  visual system 10
Tokay gecko (*Gekko gecko*) 135*f*
tongue projection 21
  biomechanics 185, 186*f*
  chameleon 110
  evolution of 105*f*
  frogs 103–4
  Hydromantes 21*f*
  motor control 106
  muscle changes 106
  rapid movements 185–6
  salamanders 106–7
  speed–accuracy trade-offs 154–5
  toads 10, 104
tortoises, temperature effects 80–1
tracking studies, predator–prey interactions 57–8, 58*f*
trade-offs 92, 143–62
  behavior 160
  expectations of 144–6
  mechanism/anatomy 92–6
  performance traits 93
  within/between species 146–7
  see also constraints; innovations

training
  aerobic 222
  animals 223–4
  endurance 222
  human athletics 222–6
traits
  fitness measure *vs.* 28
  heritability 13
  intercorrelation 26
trap-jaw ants 187–9, 188*f*, 189*f*
tree crickets, sexual selection 174
tree frogs (*Hyla crucifer*) 82
trill rate, birdsong 149–50, 150*f*
Troponin-t 203–4, 204*f*
Tucjer, Vance 60–1
turtles 52
  metabolic rate depression 77–8
  speed 52
  water diving 77–8

U
Uca pugnax (fiddler crabs) 151–2
Uma notata 76*f*
underground living 109
  morphological adaptations 191–2
Urosaurus lizards
  hindlimb length studies 38
  performance limitation studies 33–4
  sprint speed studies 38
Uta stansburiana see side-blotch lizard (*Uta stansburiana*)

V
Varanid lizards
  breathing efficiency 73, 75, 75*f*
  extreme performance 183
variability 11–13
visual system, toads 10
VO$_{2max}$
  aerobic metabolism 72–3, 72*f*
  heritability in house mice 203
  human athletes 232

mammals 73, 74*t*, 75
reptiles 73
species variation 73

W

Wainwright, Peter 215, 216*f*
wallabies, locomotion 113
Wallace, Russell 24
wall-climbing *see* adhesive
  capacities
water diving 54, 56*f*
  dolphin studies 48
  elephant seals *see* elephant
    seal diving

energetic costs 77–8
mitochondria 103*f*
turtles 77–8
  see also diving mammals
waxrunner ants, adhesive
  capacities 135
Weddel seal (*Leptonychotes
  weddellii*) 102–3
weight training 222
Weinstein, Randi 225
whale sharks (*Rhincodon
  typus*) 52, 52*f*
white muscle, distribution
  of 146

whole-organism view of
  performance 6–7,
  7–8, 7*f*
fluctuating asymmetry 177
Wilson, Alan 56–7
wing movements, insect
  flight 126–7, 127*f*
wolf spider (*Hygrolycosa
  rubrofasciata*) 174, 175*f*

Z

zebra-tailed lizard
  (*Callisaurus
  draconides*) 53–4, 55*f*

Printed in the United States
By Bookmasters